This book belongs to:

....

D1349812

Senior Editor James Mitchem
Senior Designer Elaine Hewson
Edited by Sally Beets, Hélène Hilton, Violet Peto
Designed by Karen Hood, Hannah Moore, Samantha Richiardi
Consultant Dr. Darren Naish
Producer, Pre-Production Rob Dunn
Producer John Casey
Jacket Designer Charlotte Bull
Jacket Coordinator Francesca Young
Creative Technical Support Sonia Charbonnier
Managing Editor Penny Smith
Managing Art Editor Mabel Chan
Publisher Mary Ling
Art Director Jane Bull

First published in Great Britain in 2018 by
Dorling Kindersley Limited
80 Strand, London, WC2R 0RL

A CIP catalogue record for this book
is available from the British Library.
ISBN: 978-0-2413-1617-7

Printed in Slovakia

A WORLD OF IDEAS:
SEE ALL THERE IS TO KNOW

www.dk.com

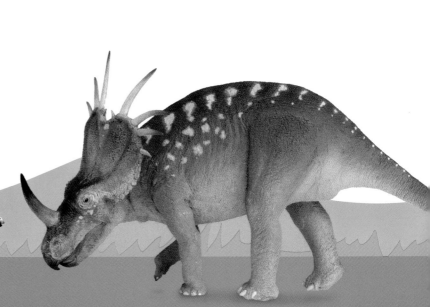

My Encyclopedia of
Very
IMPORTANT
DINOSAURS

DK

Contents

Meet the dinosaurs

Clues from the past

The **prehistoric**

Rooooooaaaarrrrr!

If you could use a time machine to visit the age of the dinosaurs, you might think that you had arrived on an alien planet. The Earth has changed so much since dinosaurs were alive, that it almost seems like a completely **different world**!

What is a dinosaur?

For 170 million years, dinosaurs were the **most dominant** land animals on Earth, but do you know what a dinosaur is?

What made a dinosaur?
There were hundreds of different dinosaurs and they came in lots of shapes and sizes, but there are some things most **had in common**.

Reptile relations
The word dinosaur means **"terrible lizard"**. Just like the lizards of today, dinosaurs were all reptiles.

Modern green basilisk lizard

Hatching from eggs
Dinosaur babies all **hatched** from eggs like birds, fish, and modern reptiles.

Prehistoric relatives

Dinosaurs lived at the same time as other creatures called pterosaurs and plesiosaurs. These animals may seem like dinosaurs, but they were totally **different** animals.

Scales

Feathers

Pterosaurs could fly, plesiosaurs could swim, but us dinosaurs ruled the land!

Claws

Claws and tails

Dinosaurs all had tails and claws. Tails helped them **balance** and claws were used for fighting or gathering food.

Scales or feathers?

Scientists used to think all dinosaurs had bodies covered in **scales**, but we now know many of them had **feathers**.

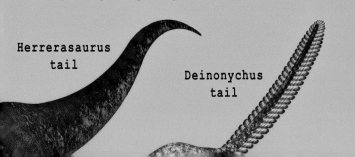

Herrerasaurus tail

Deinonychus tail

Mesozoic Era

The word "Mesozoic" means "middle animals".

The dinosaurs lived for millions of years during a time called the Mesozoic Era. The Mesozoic Era was split into **three periods**.

Brachiosaurus

Eoraptor

Compsognathus

Plateosaurus

Stegosaurus

Coelophysis

TRIASSIC PERIOD
(251–200 million years ago)

JURASSIC PERIOD
(200–145 million years ago)

Tyrannosaurus

Many of the most
well known dinosaurs
lived during the
Cretaceous period.

The time between Stegosaurus
and Tyrannosaurus was longer
than the time between
Tyrannosaurus and people!

Triceratops

Deinonychus

CRETACEOUS PERIOD
(145–66 million years ago)

PRESENT DAY
Right now!

13

The Triassic

The Triassic was the **first** of the three periods of the Mesozoic Era. It lasted for 51 million years, and was the beginning of the age of the dinosaurs.

Thecodontosaurus

The Triassic world was covered in deserts.

The world was very hot and dry during the Triassic.

Coelophysis

TRIASSIC PERIOD
251–200 million years ago

A new world

The Triassic began after a mass extinction **wiped out** almost all life on Earth. It took the planet a long time to recover, but when it did, many new types of animals appeared.

Eudimorphodon

Plateosaurus

The first dinosaurs, pterosaurs, plesiosaurs, and mammals all appeared during the Triassic.

Many insects that lived during the Triassic are still here today.

New life

The first dinosaur fossils date back around 235 million years. They show us that early dinosaurs were a lot **smaller** than the giants that followed.

Pangaea

251 million years ago, Earth's land was all connected as **one big area** called Pangaea.

The Jurassic

The second period of the Mesozoic was called the Jurassic. The mild climate created an environment for dinosaurs to **thrive**.

> Sauropods like me were the biggest land animals ever.

Apatosaurus

Archaeopteryx

Allosaurus

Stegosaurus

> I was possibly the first dinosaur that could fly!

JURASSIC PERIOD
200–145 million years ago

Changing weather

The Earth became cooler and wetter during the Jurassic. Rainfall helped new plants and lush forests to grow, which meant there was **plenty of food** for dinosaurs to eat.

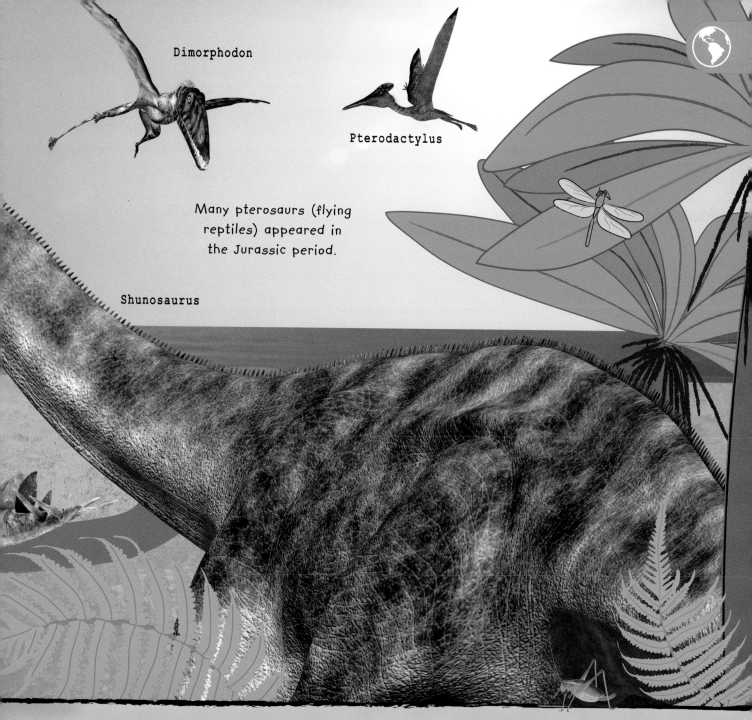

Dimorphodon

Pterodactylus

Many pterosaurs (flying reptiles) appeared in the Jurassic period.

Shunosaurus

Rise of the giants

Lots of new dinosaurs appeared during the Jurassic. Having lots of food to eat helped some, such as **sauropods**, grow to be huge!

The split

During the Jurassic, Pangaea **split** apart to make **new continents.** Oceans flooded into the spaces between them.

The Cretaceous

The third and final period of the Mesozoic was the Cretaceous. It was the **longest** of the three periods, but it didn't last forever...

Many dinosaurs developed horns and armour during the Cretaceous.

Amargasaurus

Iguanodon

Flowering plants appeared for the first time during the Cretaceous.

Triceratops

CRETACEOUS PERIOD
145–66 million years ago

The dinosaur boom

The Cretaceous was the peak of the dinosaurs time on Earth. It was during the Cretaceous that many new dinosaurs appeared, including the famous **Tyrannosaurus** and **Triceratops**.

During the Cretaceous, shallow seas flooded over the land.

Tyrannosaurus

Velociraptor

Spreading out

Earth's continents continued to drift apart during the Cretaceous. This caused dinosaurs to spread into **every corner** of the globe.

Ending on a bang

The Cretaceous ended 66 million years ago when an **asteroid** hit the Earth, killing much of the life on the planet.

A changing world

At the beginning of the age of the dinosaurs, Earth's continents (areas of land) were joined together as one big supercontinent called **Pangaea**. But over millions of years the continents split apart.

The word Pangaea means "All Earth".

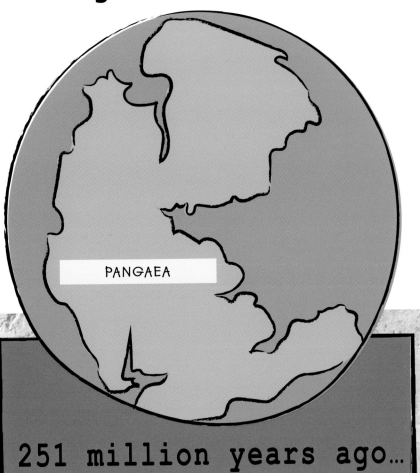

PANGAEA

Why they move
Continents move because **tectonic plates** (huge slabs of rock deep within the Earth) are slowly pushed around by the liquid rock beneath them.

251 million years ago...

At the beginning of the **Triassic**, all of Earth's continents were joined together in a C-shaped land mass called Pangaea.

200 million years ago...

During the **Jurassic**, an ocean split Pangaea into two continents called Laurasia and Gondwana, which then also began to separate.

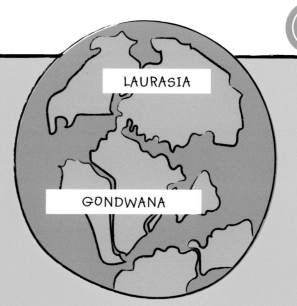

LAURASIA

GONDWANA

66 million years ago...

Over millions of years, the continents continued to drift apart. By the end of the **Cretaceous**, Earth had begun to look more like it does now.

Present day...

Today, Earth has seven continents. They are **still moving**, but only at about the speed that our fingernails grow. What might Earth look like 100 million years from now?

The Mesozoic world

If you could go back to the time of the dinosaurs, you would find a very **different** Earth to the one we live on now. The plants, weather, and most of all the animals, all changed a lot.

Warm climate

The Triassic, Jurassic, and Cretaceous all had different climates, but overall the Mesozoic was **warmer** than it is today.

There were lots of volcanic eruptions near the beginning of the Mesozoic.

Swamp

Plant life

For most of the Mesozoic, there were no flowers or grass, and the trees only had thin or spiky leaves. Some are still around, but most are **rare or extinct**.

Conifer

Fern

Cycad

Desert

Environment

At the beginning of the Mesozoic, Earth was recovering from a mass **extinction**. Much of the planet was a lifeless desert, but life bloomed over the following millions of years.

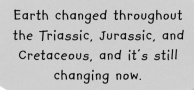

Could humans survive in the dinosaur world?

If people had been around during the Mesozoic Era, we would have struggled with the **heat** and the air would have been hard for us to breathe. But even if we could survive those conditions, we would have to find a way to hide from all the **terrifying predators**!

Earth changed throughout the Triassic, Jurassic, and Cretaceous, and it's still changing now.

When Pangaea split, new continents formed. This caused the world's landscape to change, and created new environments such as forests, rivers, mountains, and swamps.

Types of dinosaur

There were so many types of dinosaur that it can be hard to keep track of them. Luckily, experts split them into **groups** and subgroups. Here are some of the main ones.

Suchomimus

A story of hips

Experts used to think that dinosaurs could be put into groups based on the **shape of their hips**, but new evidence suggests this probably isn't the case after all.

Theropods (THERRO-pods)

These dinosaurs ranged in size, from the terrifying Tyrannosaurus to speedy Velociraptor. They all walked on two legs and ate meat.

Tyrannosaurus

Velociraptor

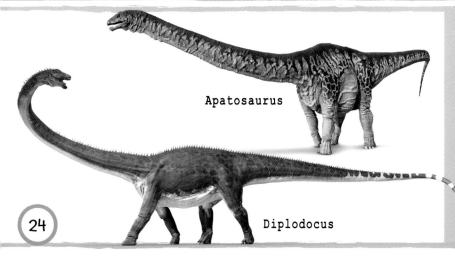

Apatosaurus

Diplodocus

Sauropodomorphs (SORE-oh-POD-oh-morfs)

These enormous plant-eaters were the largest creatures to ever walk the Earth. They had long necks and tails.

Thyreophorans (THIGH-ree-OFF-oh-rans)

These plant-eaters walked on four legs and had protective spikes and armoured plates.

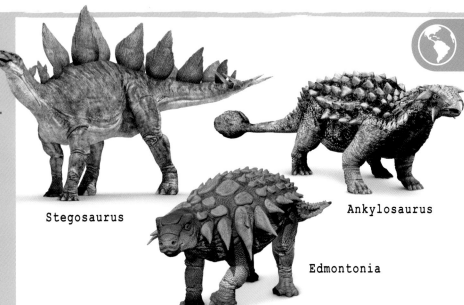

Stegosaurus

Edmontonia

Ankylosaurus

Ornithopods (OR-nith-oh-pods)

A common group of dinosaurs, ornithopods were browsers who walked on two legs and sometimes travelled in groups.

Parasaurolophus

Iguanodon

Hypsilophodon

Marginocephalians (MAR-jee-no-sa-FAY-lee-ans)

Common in the Cretaceous period, these dinosaurs had bony frills on their heads. Some walked on two legs, others walked on four.

Triceratops

Pachycephalosaurus

Chasmosaurus

What did dinosaurs look like?

Dinosaurs could be big, tiny, colourful, feathered, or scaly. There's no such thing as a "normal" dinosaur!

Compsognathus

Giganotosaurus

Some dinosaurs were really massive, but others, such as Compsognathus, were only about the size of a chicken.

Lookalikes

These amazing animals came in different shapes, sizes, and probably colours. Each type was unique but they did have some things in **common**.

 Dinosaurs had scales or feathers. Some had both.

 All dinosaurs walked upright on legs and had claws on their fingers and toes.

 Every dinosaur had a tail.

Looking the part

Dinosaurs could look very different from each other, but some dinosaurs had interesting features that really made them stand out.

Spinosaurus had an enormous sail on its back.

Caudipteryx had soft feathers all over its body.

Parasaurolophus had a curved crest on its head.

Giant sauropods like Diplodocus had very long necks and tails.

Cryolophosaurus had a little crest on top of its head. It was first named "Elvisaurus".

Kentrosaurus had large bony plates running along its back and tail.

Triceratops was famous for its three horns and impressive neck frill.

What **else** was there?

Dinosaurs are amazing, so it's no wonder they get a lot of attention, but there were plenty of **other animals** alive at the same time.

Beelzebufo was the size of a beach ball!

The first insects appeared over 400 million years ago!

Beelzebufo

Insects

Some insects such as dragonflies, cockroaches, and millipedes were alive **before** the dinosaurs, and others such as ants, flies, and bees appeared throughout the Mesozoic.

Fish

Long before dinosaurs, the world's oceans were **full of life**. Plenty of reptiles, shellfish, and fish – including sharks – already existed.

Amphibians

Amphibians appeared before the dinosaurs, and have been around ever since. One ancient frog, Beelzebufo, was so **big** it might have eaten small dinosaurs!

Humans are mammals, but we weren't around when dinosaurs lived.

Pterosaurs were flying reptiles and plesiosaurs were sea-dwelling reptiles.

Mammals

Today many of Earth's mammals, such as whales and elephants, can grow to be huge. But the mammals of the Mesozoic, like Eomaia, were much **smaller**.

Eomaia

Reptiles

Dinosaurs were prehistoric reptiles, but **not all** prehistoric reptiles were dinosaurs. Early crocodiles, snakes, and turtles, as well as pterosaurs and plesiosaurs, all lived alongside dinosaurs.

Some prehistoric turtles were as big as a family car.

They aren't dinosaurs!

The creatures that dominated the prehistoric **skies** and **seas** might have looked like dinosaurs, but they were totally different animals.

Ornithocheirus

Pterosaurs

These winged wonders were prehistoric **flying reptiles**. They had light bodies, and their wings were made of stretched skin that was attached to their bones.

Rhamphorhynchus

We ate plenty of fish, but we also scavenged for a lot of our meals.

Pterodactylus

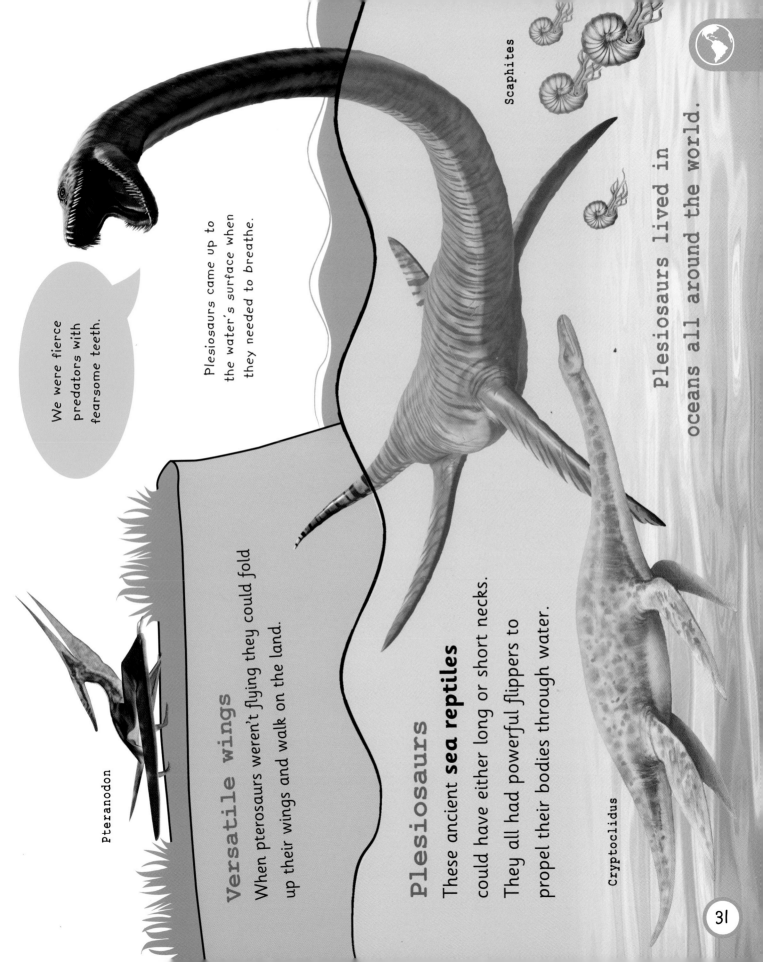

We were fierce predators with fearsome teeth.

Plesiosaurs came up to the water's surface when they needed to breathe.

Scaphites

Plesiosaurs lived in oceans all around the world.

Pteranodon

Versatile wings

When pterosaurs weren't flying they could fold up their wings and walk on the land.

Plesiosaurs

These ancient **sea reptiles** could have either long or short necks. They all had powerful flippers to propel their bodies through water.

Cryptoclidus

Ancient insects

Dinosaurs can be so big that it's easy to overlook the little creatures. But just like today, **insects** were buzzing around in the Mesozoic Era.

The very first insects

Dragonflies and cockroaches appeared before the dinosaurs, around 300 million years ago. They are still around today.

Some dinosaurs would have eaten insects like us.

Ant

Bee

Ants	Bees
Colonies of ants first appeared 100 million years ago. Today, there are around **12,000** different types.	Bees appeared during the Early Cretaceous, a few million years after the first **flowering plants**.

All about insects

Insects are the most widespread, successful, and varied group of creatures on Earth. They have several things in common.

✓ **All insects have six legs.**

✓ **Their bodies have three sections.**

✓ **Most insects have wings.**

Dragonfly

Butterfly

Fly

Flies

Flies appeared in the Triassic period, around the **same time** as the first dinosaurs.

Butterflies

The first butterflies appeared around 66 million years ago, near the time that the dinosaurs **died out**.

Ancient fish

Fish had been swimming about in the waters of our world **long before** the dinosaurs appeared.

Spinosaurus

Hybodus

Leedsichthys

Coelacanth

Coelacanths are sometimes called "living fossils".

Hybodus

This ancient **shark** had two sets of teeth: one very sharp and the other blunt. It went extinct in the Cretaceous period.

Leedsichthys

This bony Jurassic fish was **truly massive**. Leedsichthys was about the same size as a modern killer whale.

Coelacanth

Scientists thought this fish died out with the dinosaurs, but amazingly it was **rediscovered** in 1938!

What are fish?

Fish are a large group of animals. There are thousands of different types, with lots in common.

 They use gills to breathe underwater.

 They can be found in fresh or saltwater.

 Most of them have scales.

 They have fins to steer and stay upright in water.

Some dinosaurs, such as Spinosaurus, feasted on a steady diet of fish.

Sawfish

Lepidotes

Fossils of Lepidotes have been found all over the world. This fish was one of **Baryonyx's** favourite meals.

Lepidotes fossil

Sawfish

Sawfish are nicknamed "carpenter sharks" because of their long, saw-like nose extensions. These fish appeared in the Cretaceous and are still around **today**.

Ancient reptiles

Reptiles definitely **ruled** the Mesozoic Era. But dinosaurs, pterosaurs, and plesiosaurs weren't the only reptiles around at the time.

Postosuchus

Archelon

Dakosaurus

Deinosuchus had a bite as powerful as Tyrannosaurus!

Deinosuchus

Archelon

Fossils of turtles date back as far as 220 million years ago. At 4m (13ft), Archelon is one of the **biggest** turtles ever.

Deinosuchus

The mighty Deinosuchus was one of the most powerful predators of the Cretaceous. It was twice as big as a modern **alligator**.

Postosuchus

Although it looked like a dinosaur, Postosuchus was actually related to **crocodiles**. It probably ate early dinosaurs.

What are reptiles?

Snakes, lizards, crocodiles, and tortoises are all reptiles. They have several things in common.

 Their bodies are covered in waterproof scales.

 They are cold-blooded and need the Sun to warm up.

 They all hatch from eggs.

My name means "Earth lizard" because scientists used to think that I lived on land.

Geosaurus

Desmatosuchus

Dakosaurus

This marine predator roamed the oceans of the Jurassic and Cretaceous. It was a speedy swimmer with a **brutal** bite.

Geosaurus

Geosaurus was a strange-looking predator that swam around the Jurassic and Cretaceous oceans **hunting** for fish.

Desmatosuchus

This Triassic reptile was a relative of the modern crocodile. Desmatosuchus had **bony** plates along its back and tail.

Ancient mammals

Early mammals lived in a world ruled by reptiles, so life wasn't easy. The mammals of the Mesozoic were **very different** to modern ones.

> The first mammals were all tiny, but today, the biggest animals in the world are mammals.

Nemegtbaatar

Megazostrodon

Morganucodon

Nemegtbaatar

Although this mammal looks a lot like a rat, they're not closely related. It had large front teeth and a very wide **snout**.

Megazostrodon

This tiny, furry animal probably ran, climbed, and burrowed just like a modern day **rat**.

Morganucodon

This little insect-eater was one of the **first** mammals to ever exist. It appeared in the Late Triassic.

Morganucodon jaw bone

What are mammals?

Mammals are a group of animals that come in all shapes and sizes. You're a mammal too! There are a few things we have in common.

 Most mammals are born, rather than hatch from eggs.

 They have bony skeletons.

 Their bodies are hairy.

 They are warm-blooded.

Sinoconodon

Most early mammals ate insects and other little animals.

Eomaia

Eomaia

Only one Eomaia fossil has ever been found but it was very well preserved. It shows that Eomaia had little hands and feet that were great for **climbing**.

Sinoconodon

Although it could have sat in the palm of your hand, Sinoconodon was one of the **biggest** mammals of the Jurassic period.

Zalambdalestes

Zalambdalestes

This cute creature had unusually **long** back legs. It might have moved by hopping around.

Life as a **dinosaur**

Even though they were the most dominant animals of the time, life **wasn't always easy** for dinosaurs. It could be a challenge to find enough food to eat and avoid being eaten, so each dinosaur needed special skills to stay alive.

Dino dinner

Based on what they **ate**, dinosaurs can be split into two main categories: meat-eaters and plant-eaters. Some dinosaurs ate both, but most usually stuck to one or the other.

MEAT MENU

Other dinosaurs

Small mammals

Reptiles

Fish

Insects

Eggs

Tyrannosaurus

I'd better run if I don't want to be eaten!

Meat-eaters

Dinosaurs that ate meat were **carnivores**. Most had good eyesight, long legs, and sharp teeth to help them hunt, but others scavenged for their meals.

PLANT MENU

Leaves

Mosses

Berries

Seeds

Fruit

OMNIVORE MENU

Plants

Small mammals

Insects

Lizards

Fruit

I'm so big I needed to eat LOTS of plants.

Plant-eaters

These dinosaurs usually had blunt or flat teeth to help them strip and chew plants. They were called **herbivores**.

Both!

Some dinosaurs ate both meat and plants. The diet of these **omnivores** was made up of plants and fruit, as well as lizards, fish, and insects.

We had both flat and sharp teeth, so scientists think we were probably omnivores.

Maiasaura

Some dinosaurs were "accidental omnivores" because they would have eaten tiny insects on leaves without realizing.

Heterodontosaurus

43

Meat-eaters

Hunting in the Mesozoic required very special **skills** and weapons. Meat-eating carnivores had to be seriously fierce.

Carnivore tools

Strong jaws

Powerful jaws were ideal for grabbing prey and crushing their bones.

Sharp teeth

Jagged teeth helped to pierce and rip meat.

Deadly claws

Carnivores used their claws to slash or grab prey.

Fast legs

Strong legs helped hunters chase down speedy prey.

Good eyesight

Forward-facing eyes gave carnivores excellent vision to spot prey.

Thick tails

A chunky tail helped meat-eaters keep their balance.

Specialized tools for special diets

Compsognathus

Tyrannosaurus

Spinosaurus

Plant-eaters

Most dinosaurs ate plants, but different herbivores had special **tools** that helped them eat different plants.

Herbivore tools:

Strong beaks

Ceratopsians had sharp beaks that helped them rip off tough pine and fern plants.

Flat teeth

Beaked dinosaurs often had flat teeth to chew their food.

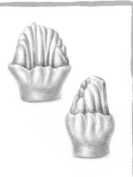

Long necks

Sauropods had long necks that let them reach the leaves high in the treetops.

Peg teeth

Long-necked herbivores had front teeth to cut and grab leaves before swallowing them.

Strong mouths

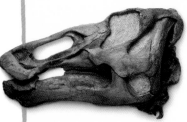

Hadrosaurs could clamp their mouths around branches to strip off lots of leaves at once.

Lots of teeth

Duck-billed dinosaurs had up to a thousand teeth for grinding food. That's a lot more than most dinosaurs.

Triceratops

This famous ceratopsian could eat very tough plants using its sharp **beak**. It also had strong teeth to help grind them down.

Brachiosaurus

Giant sauropods like Brachiosaurus ate **without chewing**. It may have swallowed stones to grind up the plants inside its stomach instead.

Edmontosaurus

This hadrosaur wasn't really able to defend itself from predators, so it used its many teeth to **eat quickly**, so it didn't stay in the same place for long.

Dinosaur homes

Dinosaurs lived in different places all over the world. The environments they lived in are called **habitats**. Here are some of the main ones.

Rivers and coasts

Lots of plants grow near water, so there was a lot of **food** for plant-eaters by rivers and coastlines – as well as fish for meat-eaters.

Look at all these plants to eat.

Deserts

Some resourceful dinosaurs were able to survive in hot deserts by eating **special plants** as a source of food and water.

Desert life can be tough.

Plains

Open plains and scrublands were perfect places for herds of dinosaurs to **roam** across in search of food.

It was hard for predators to sneak up on me on flat plains.

Woodlands

Dinosaurs liked forests because there was so much to eat. Some woodlands were **cool** and others were hot and **swampy**.

The Earth has changed a lot since dinosaurs existed. Some fossils found on beaches were from dinosaurs that lived in the mountains.

While many dinosaurs lived by water, none of them lived in it. Other reptiles such as **plesiosaurs** ruled the oceans.

49

Sticking together

Many dinosaurs lived and travelled in groups called **herds**. By sticking together, dinosaurs could help and look out for each other.

Some herbivores travelled together to find fresh food for the herd to eat.

Einiosaurus

Fossil friends

One reason palaeontologists believe some dinosaurs lived in herds is because fossil **"trackways"** show the footprints of many dinosaurs walking together.

Fossil footprints

Safety in numbers

The main reason dinosaurs formed herds was to **protect themselves** from predators. Not only were they more dangerous as a group, but they could also alert each other if danger was nearby.

Fossils show that herds may have kept their young in the centre of the group to shield them from attacks.

Hungry hunters

Clever predators, such as Velociraptor, may have **worked together** to hunt larger or more dangerous dinosaurs.

Some modern wild animals, such as elephants and zebras, live in herds, too.

Dino communication

Dinosaurs needed to attract mates and warn each other about danger. They couldn't have conversations, but they did have other ways of **communicating**.

Weapon warnings

Many dinosaurs wanted to **avoid fights** if possible. Showing off their impressive weapons or armour was a great way to warn predators that it might be risky to attack them.

Kentrosaurus

Don't eat me! I'm sure you can find a less spiky meal.

Frills

Although it was only made of light bone and stretched skin, **Torosaurus'** neck frill still looked very intimidating.

Torosaurus

Spikes

Dinosaurs such as **Kentrosaurus** had huge bony spikes sprouting from their heads, shoulders, and tails. Carnivores might think twice before attacking such a spiky herbivore.

Fossils give us lots of clues about dinosaur

HOOT?
HOOT?

Noisy dinos

It's possible that some dinosaurs would have been able to make **calls** and other **noises**. But it's hard to know exactly what kind of sounds they made.

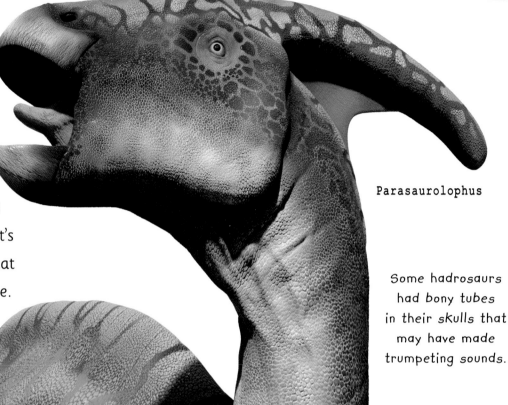

Parasaurolophus

Some hadrosaurs had bony tubes in their skulls that may have made trumpeting sounds.

Sails

Predators picked off small, weak dinosaurs to eat. **Ouranosaurus** had a sail on its back that made it look bigger than it was.

Ouranosaurus

Crests

It would have been useless in a fight, but **Lambeosaurus'** head crest was great for recognizing other members of the herd.

Lambeosaurus

communication, but there is still a lot we don't know.

Dinos for **dinner**

Predatory dinosaurs were the deadliest animals to ever walk the Earth, but hunting was dangerous, and they needed **weapons and skills** to get the job done.

Deinonychus

Brains over brawn

Some carnivores, such as Deinonychus, were **clever**, and had well-developed brains and senses. They may have hunted in packs to trap their prey.

Special weapons

Spinosaurus **specialized** in hunting one type of prey. It had long, narrow jaws full of cone-shaped teeth to grip onto wet, wriggling fish.

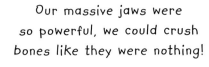

Our massive jaws were so powerful, we could crush bones like they were nothing!

Tyrannosaurus

Spinosaurus

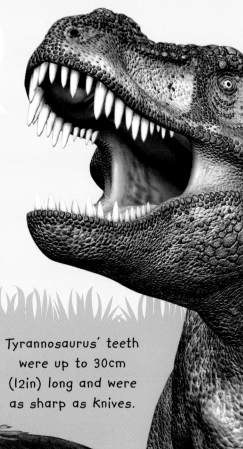

Tyrannosaurus' teeth were up to 30cm (12in) long and were as sharp as knives.

Microraptor

Flying high

Little meat-eaters didn't stand a chance against big prey. Some, such as Microraptor, evolved to fly or glide so that they could **swoop** down and snatch small animals.

The big ones

Giant theropods, such as Tyrannosaurus, were **apex predators**. This means they were so strong and deadly they had nothing to fear except each other.

Fight or... run!

What happened when a dinosaur came face to face with a predator? There were only a few options...

Giants like Brontosaurus were rarely attacked because they were so big, but sometimes predators got desperate.

Grrrrr!

Allosaurus

Look scary

Some herbivores, such as Pentaceratops with its sharp horns, looked very intimidating. A predator might decide attacking wasn't **worth the risk**, and look for easier prey instead.

Strong horns

THWACK!

Brontosaurus

Attacking a giant dinosaur could be very dangerous.

Run away

Small, defenseless dinosaurs had no choice except to try to **outrun** predators. Luckily, the fastest ones could run at 80kph (50mph)!

Strike back

Ankylosaurus was covered in protective armour, but it also had a big **club** on its tail to smash enemies that got close.

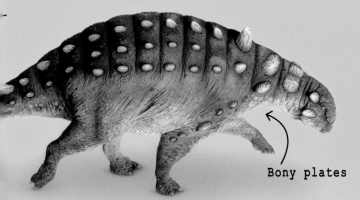

Bony plates

Fight

Some dinosaurs had useful **defensive weapons**. Therizinosaurus' claws were mainly used to grab branches, but they would have also been useful in a fight.

Sharp claws

57

Dinosaur eggs

Just like modern reptiles, dinosaurs **hatched from eggs**. Fossilized eggs can help us understand how young dinosaurs developed.

Some dinosaurs cared for their young after they hatched, but others were left to fend for themselves.

Citipati

Nesting

Some dinosaurs, such as **Citipati**, built nests for their eggs and sat on them to keep them warm like many birds do today.

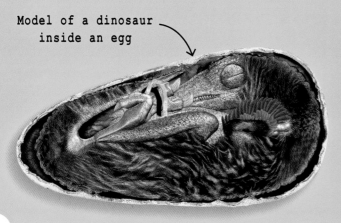

Model of a dinosaur inside an egg

Small eggs, big dinosaurs

Dinosaur eggs varied in size and shape depending on the dinosaur, but they were all **relatively small** in comparison to a fully grown dinosaur. Experts believe this means dinosaurs probably grew to full size quickly.

Maiasaura
nest

Apatosaurus
egg

Safety in numbers

A huge Maiasaura nesting site found in the USA in 1979 proved that some dinosaurs built their nests close to **each other**. They probably did this for safety.

How big were dinosaur eggs?

Oviraptor
philoceratops
egg

Chicken
egg

An Apatosaurus egg
was about the size
of a basketball!

Meet the

dinosaurs

From massive meat-eaters and giant grazers, to walking tanks and mini monsters, dinosaurs came in more **shapes and sizes** than you might think. Dig through these pages to learn about these incredible animals and their prehistoric relatives.

Maiasaura

Plateosaurus

Heterodontosaurus

Parasaurolophus

Psittacosaurus

Meet the grazers

These chilled out herbivores spent their days **munching on plants** – and lots of them! They weren't the biggest or most fierce dinosaurs, so some of them travelled in herds to keep safe.

Iguanodon

Heterodontosaurus
HET-er-oh-DON-toe-SORE-us

This speedy little dinosaur was only about the size of a turkey. Its **unusual teeth** set it apart from most other dinosaurs.

> Most dinosaurs only had one type of tooth, but we had several.

What a mouthful

Heterodontosaurus' mouths had small teeth in the front, sharp teeth on the sides, and a beak. It could choose the **best way** to eat depending on its food. It also had long fangs.

Five curved fingers

251 million years ago	200		145	66
Triassic		Jurassic	Cretaceous	

Fact File

Our long fangs were probably used for **FIGHTING** instead of **EATING**.

Small but fast
Strong **hind legs** helped Heterodontosaurus quickly run away from attackers.

Heterodontosaurus means "different-toothed lizard".

This Heterodontosaurus fossil from South Africa is one of the most complete fossils ever found.

Size: 1m (3ft) long   **Habitat:** Scrubland   **Diet:** Plants and insects

Psittacosaurus
si-tak-ah-SORE-us

This little dinosaur was a relative of the massive horned Triceratops. They were part of the **ceratopsian** family.

We had small, spiky horns that stuck out from our faces.

PARROT

Horn

Big beak
Psittacosaurus was named "parrot lizard" because of its bird-like **beak**. Although it also had teeth, its strong beak would have been useful to crack open seeds and nuts it found on the forest floor.

251 million years ago · 200 · 145 · 66

Triassic · Jurassic · Cretaceous

Fact File

We were among the oldest and smallest members of the ceratopsian family.

Bristles

Scaly body

Bristles

Fantastic fossils

Scientists have discovered lots of Psittacosaurus fossils so they know quite a lot about it. This fossil clearly shows it had **bristles** along its tail.

Long hind legs

Size: 2m (6½ft) long   **Habitat:** Forests   **Diet:** Plants and seeds

Iguanodon

ig-WAH-no-don

This dinosaur was mostly interested in munching plants, but it had a **secret weapon** to defend itself when needed.

Fossilized thumb spike

Secret weapon

The sharp thumb **spikes** on Iguanodon's hands were probably used to reach tall branches, and they would also have been useful as defense from predators.

251 million years ago	200	145		66
Triassic	Jurassic		Cretaceous	

Fact File

Scientists used to think our thumb spike was a face horn, a bit like a rhino's.

A group of 38 Iguanodon skeletons were found together in Belgium, so they may have lived in **HERDS**.

We were about as long as a bus and weighed as much as an elephant.

An early discovery

Iguanodon was the first plant-eating dinosaur ever discovered. It got its name because its teeth looked like they belonged to a giant **iguana**.

The thumb spike was about 14cm (6in) long.

Size: 9m (30ft) long　　**Habitat:** Forests　　**Diet:** Plants

Plateosaurus
plate-ee-oh-SORE-us

This early dinosaur was a **prosauropod**, an ancestor of giant sauropods such as Diplodocus. But unlike sauropods, plateosaurus only walked on two legs.

Germany

Lots of Plateosaurus skeletons were discovered in the same place in Germany.

251 million years ago	200	145	66
Triassic	Jurassic	Cretaceous	

Fact File

Strong grazer

Although it was heavy, Plateosaurus could probably **run quite quickly**. It used its hands to grab food and its strong teeth helped it chew tough leaves.

The word prosauropod means "before sauropods".

Plateosaurus had a long flexible neck that helped it feed from tall trees.

Strong tail for balance

A sticky end

Many Plateosaurus skeletons have been found together. Scientists think a herd of them became **stuck in mud** and sank. The mud preserved their bodies as fossils for millions of years.

Size: 8m (25ft) long   **Habitat:** Deserts   **Diet:** Plants

Hypsilophodon
hip-sih-LOAF-oh-don

This speedy little **plant-eater** was the perfect size to dash through the undergrowth and **stay hidden** from bigger predators.

Our powerful legs helped us run quickly, and our stiff tails kept us balanced.

Hypsilophodon had four fingers and a thumb on each hand, just like people.

Stiff tail

251 million years ago	200	145	66
Triassic	Jurassic	Cretaceous	

 **Fact File**

Leaf shredder

Hypsilophodon had a **sharp, beak-like mouth** that was ideal for grabbing and tearing plants. Special teeth at the back of its mouth helped it mash up the plants.

Fossilized skull

Hypsilophodon had very big eyes and great vision. It might have been active at night.

Narrow beak

Hypsilophodon herds

Lots of fossilized Hypsilophodon footprints have been found close together, so some experts think Hypsilophodon lived in **herds**.

Size: 2m (7ft) long   **Habitat:** Forests   **Diet:** Plants

Parasaurolophus

PA-ra-SORE-oh-LOAF-uss

Sometimes called the dinosaur version of a cow, Parasaurolophus belonged to a group of dinosaurs called hadrosaurids, which means "**duck-billed lizards**".

> Unlike most herbivores, we could walk on either two or four legs.

Trumpet crests

Parasaurolophus had a crest on its head that may have helped identify males and females. Many experts also believe its crest could be used to make **sounds**.

Parasaurolophus lived and traveled in large herds.

251 million years ago   200   145   66

Triassic   Jurassic   Cretaceous

Fact File

Crest

We had hundreds of **GRINDING TEETH** to mash up all the plants we needed to eat.

Size: 9m (30ft) long   **Habitat:** Woodlands   **Diet:** Plants

Maiasaura

MY-ah-SORE-a

This herbivore lived in **big groups** that probably raised their babies together. That's why Maiasaura means "good mother lizard".

At hatching, we were about 30cm (12in) long (about the size of this book), but we grew to be huge!

Safety in numbers

Maiasaura had no defensive weapons to fight off predators, but it lived in **herds** made up of hundreds of dinosaurs. Sticking together meant the herd could look out for each other.

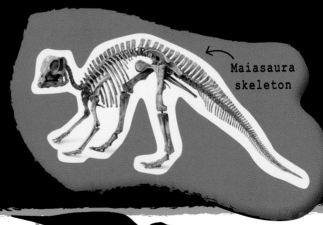

Maiasaura skeleton

251 million years ago	200	145	66
Triassic	Jurassic	Cretaceous	

Fact File

Happy family

Maiasaura dug nests in the ground for its eggs and covered the eggs with leaves to keep them warm. Once the eggs hatched, the babies **stayed in the nest** and were fed until they were strong enough to feed themselves.

Even after leaving the nest, young Maiasaura kept close to their mothers while they were growing up.

Baby
Maiasaura

In 1985 astronauts took a Maiasaura fossil into space!

Size: 9m (30ft) long   **Habitat:** Plains   **Diet:** Plants   77

Coelophysis

Allosaurus

Deinonychus

Utahraptor

Velociraptor

Meet the predators

Just like today, many animals from the Mesozoic Era were skilled **hunters**. But these prehistoric carnivores are now the stuff of legend. Do you dare turn the page and read about these fearsome creatures?

Allosaurus

al-oh-SORE-uss

This terrifying hunter had speed, deadly claws, and dagger-sharp teeth – all the tools it needed to be one of the **fiercest predators** of the Jurassic period.

Some experts think we used our teeth to **SLASH** at our prey.

Powerful legs gave Allosaurus the speed to chase down prey.

251 million years ago   200   145   66

Triassic   Jurassic   Cretaceous

Fact File

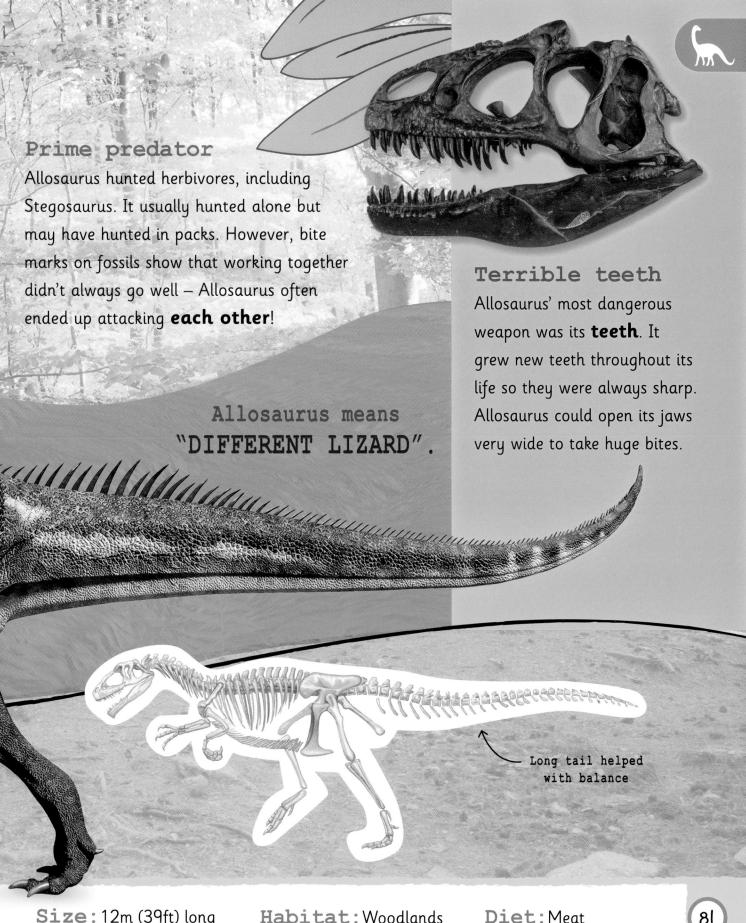

Prime predator

Allosaurus hunted herbivores, including Stegosaurus. It usually hunted alone but may have hunted in packs. However, bite marks on fossils show that working together didn't always go well – Allosaurus often ended up attacking **each other**!

Allosaurus means
"DIFFERENT LIZARD".

Terrible teeth

Allosaurus' most dangerous weapon was its **teeth**. It grew new teeth throughout its life so they were always sharp. Allosaurus could open its jaws very wide to take huge bites.

Long tail helped
with balance

Size: 12m (39ft) long   **Habitat:** Woodlands   **Diet:** Meat

Velociraptor

vel-OSS-ee-rap-tor

This feathered predator may have been small, but it was fast, vicious, and may have worked as a **team** to hunt its prey.

Sharp claws

Feathers

Velociraptor means "speedy thief".

251 million years ago

200

145

66

Triassic

Jurassic

Cretaceous

Fact File

We might have hunted big prey as a pack, jumping onto its back and slashing with our claws.

Foot claw

Final fight
Scientists have found an incredible fossil of a Velociraptor attacking a Protoceratops by pinning it down with its **foot claw**. Both dinosaurs died from their wounds and were buried in sand.

Small but deadly
Despite its size, Velociraptor was a **dangerous** hunter. Once it caught its prey, it used the sharp claws on its hands and feet, as well as its razor-sharp teeth, to take it down.

Size: 2m (6½ft) long　　**Habitat:** Deserts　　**Diet:** Meat

Coelophysis
SEE-lo-FYE-sis

This sleek, speedy hunter from the Late Triassic was scary enough on its own. But a **pack** of them? That's a terrifying thought!

We had three long fingers and one short one on each hand.

Short and long fingers

Animal remains in stomach

One Coelophysis fossil was in such good condition that scientists could see what it had eaten the day it died – a small reptile called Hesperosuchus.

251 million years ago	200	145	66
Triassic	Jurassic	Cretaceous	

Fact File

Long neck

Made to hunt

Coelophysis was an excellent hunter with good eyesight and the ability to run very fast. Its teeth were small but shaped like **hooks** – good for catching prey.

Teamwork

Scientists think that Coelophysis probably lived and hunted in **family groups,** because hundreds of its skeletons were found in the same place in New Mexico, USA.

Thin tail

Adult Coelophysis may have taught the young ones how to hunt.

Size: 3m (10ft) long   **Habitat:** Desert plains   **Diet:** Small animals

Suchomimus

SOO-ko-MIME-us

With its powerful body and long jaws, the fearsome Tyrannosaurus-sized predator Suchomimus was perfectly suited to **terrorize fish** in its swampy home.

Crocodile look-a-like

Suchomimus means "crocodile mimic". Like a crocodile, Suchomimus had **long, narrow jaws** and sharp teeth. It's possible it paddled in water and poked its snout out to breathe.

Suchomimus may have

Suchomimus fossils have been found in the Sahara desert in Africa. Millions of years ago the Sahara was a swamp.

251 million years ago   200   145   66

Triassic   Jurassic   Cretaceous

Fact File

had spikes along its back and tail.

We might have **GRABBED FISH** with our powerful arms.

Eek!

Built to fish

Suchomimus had a perfect mouth for fishing. It had more than 120 teeth, which pointed toward the **back** of its mouth to help trap slippery prey.

Size: 9m (30ft) long   **Habitat:** Swamps   **Diet:** Fish and marine reptiles

Utahraptor
YOU-tah-RAP-tor

This feathered hunter was very fast and deadly. It used its **huge toe claws** to easily bring down its prey.

These hunters might have teamed up to take on bigger dinosaurs.

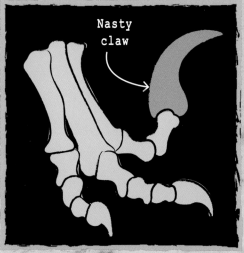

Nasty claw

Deadly weapon

Once it had chased down its target, Utahraptor would slash or stab it with its sharp toe claws. These claws were almost the size of **this page**!

251 million years ago   200   145   66

Triassic   Jurassic   Cretaceous

Fact File

Utahraptor was big and fast. It weighed about the same as a **POLAR BEAR**.

The word raptor means "thief".

Super raptor

Utahraptor belonged to a group of dinosaurs called **dromaeosaurs**. All the dromaeosaurs were vicious hunters, but Utahraptor was the biggest and most fierce.

Toe claw

Size: 7m (23ft) long　　　**Habitat:** Plains　　　**Diet:** Meat

Deinonychus
dye-NON-ee-cuss

It wasn't the biggest predator, but a dinosaur whose name meant **"terrible claw"** was clearly not to be messed with!

> We took advantage of our great speed when hunting.

Pack attack

It's possible that Deinonychus sometimes **hunted as a pack**. Scientists think the pack took on bigger prey by jumping on its back and attacking it with their claws and teeth.

251 million years ago	200	145	66
Triassic	Jurassic	Cretaceous	

Fact File

My claws weren't my only weapons. I had 60 **SHARP** teeth, too.

Deadly claw

THAT claw

Not only did Deinonychus have sharp claws on its hands, it had an especially scary claw on each of its feet. This "terrible claw" could inflict **nasty wounds** on its target.

Size: 3m (10ft)   **Habitat:** Forests and swamps   **Diet:** Meat

91

Latenivenatrix

lah-ten-EYE-vuh-NAY-tricks

Once known as **Troodon**, Latenivenatrix was not the biggest or strongest dinosaur, but it got by on its intelligence, sharp eyesight, and quick feet.

Built for speed

As a member of the bird-like dinosaur group troodontids, Latenivenatrix was **quick on its feet**. Its light body and long legs helped it to run away from predators and chase prey.

Latenivenatrix raced through woodlands looking for food.

251 million years ago   200   145   66

Triassic   Jurassic   Cretaceous

Fact File

Excellent eyesight

Instead of having side-facing eyes like most dinosaurs, Latenivenatrix's eyes **faced forward**. This helped it to judge distances when it hunted.

I had one of the biggest brains in relation to body size of any dinosaur (but it still wasn't very big!).

Strong, slender legs

Visible feathers

Natural nesters?

Adult Latenivenatrix fossils have been found on top of nests containing eggs. This means that Latenivenatrix probably **sat on its eggs** like many birds do today.

Size: 3m (10ft)　　**Habitat:** Woodlands　　**Diet:** Small animals

Herrerasaurus

heh-RARE-ra-SORE-uss

This swift predator from the middle of the Triassic was one of the very **first** dinosaurs to have existed.

We had a flexible lower jaw that helped us cling to our prey.

Early hunter

The early dinosaurs were smaller than many of the giants that would come later. Scientists think Herrerasaurus might have been an early **theropod**, but new evidence suggests that may be wrong.

251 million years ago

200

145

66

Triassic   Jurassic   Cretaceous

Fact File

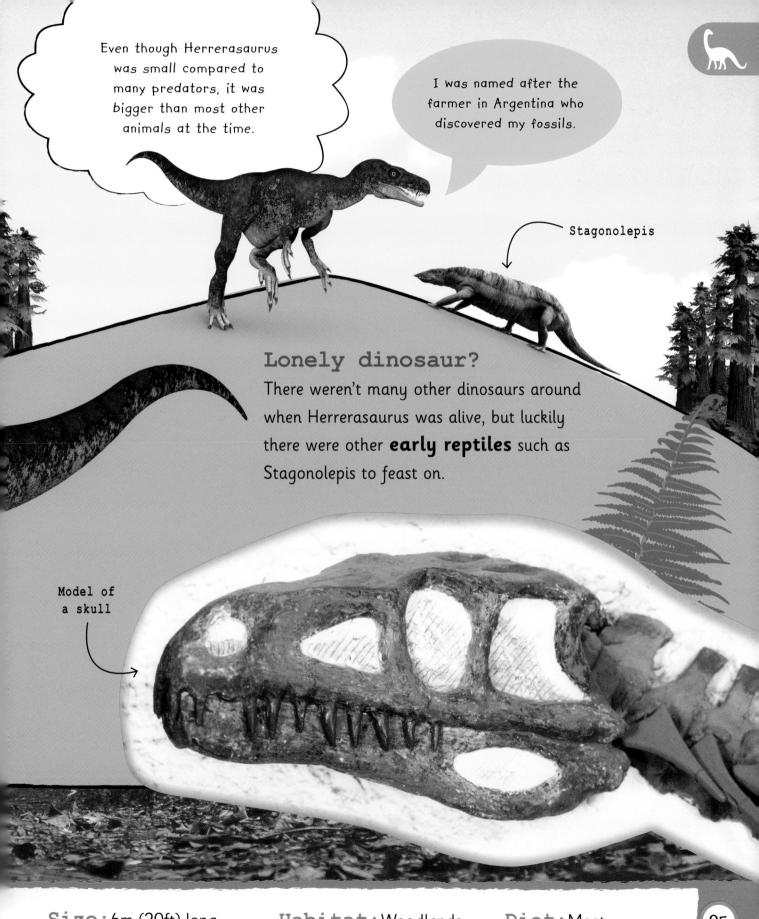

Even though Herrerasaurus was small compared to many predators, it was bigger than most other animals at the time.

I was named after the farmer in Argentina who discovered my fossils.

Stagonolepis

Lonely dinosaur?

There weren't many other dinosaurs around when Herrerasaurus was alive, but luckily there were other **early reptiles** such as Stagonolepis to feast on.

Model of a skull

Size: 6m (20ft) long   **Habitat:** Woodlands   **Diet:** Meat

Tyrannosaurus
tie-RAN-oh-SORE-us

Nicknamed the **king** of dinosaurs, Tyrannosaurus ruled the Cretaceous. It was the most powerful land predator ever, and no animal was a match for it.

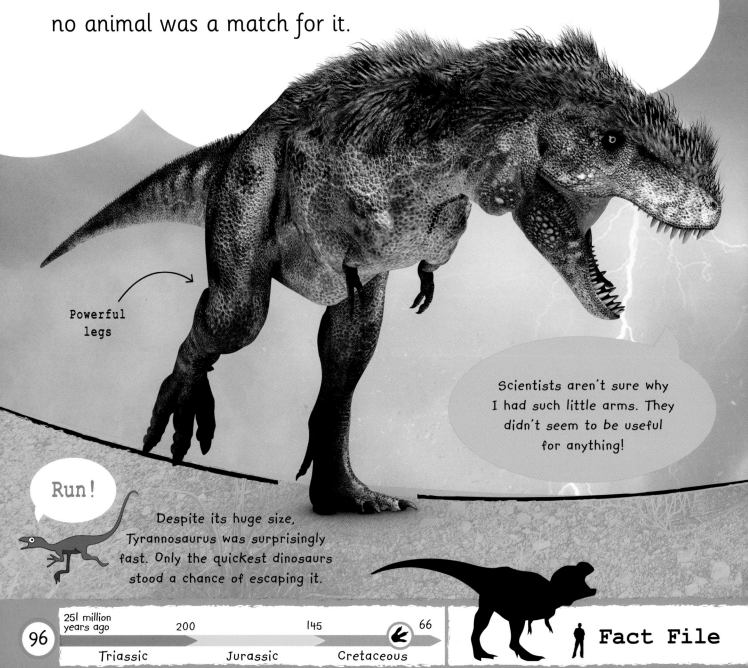

Powerful legs

Scientists aren't sure why I had such little arms. They didn't seem to be useful for anything!

Run!

Despite its huge size, Tyrannosaurus was surprisingly fast. Only the quickest dinosaurs stood a chance of escaping it.

251 million years ago	200	145	66
Triassic	Jurassic	Cretaceous	

Fact File

I'm the dinosaur

king!

Tyrannosaurus had the most powerful bite of all time. Its teeth could easily crunch through bones.

Massive footprint fossil

Claim to fame

It's no wonder Tyrannosaurus is one of the most famous dinosaurs. It was a mega monster that didn't have to be **afraid of anything** except for another, bigger Tyrannosaurus!

Size: 12m (39ft)   **Habitat:** Forests and swamps   **Diet:** Meat

97

Diplodocus

Giganotosaurus

Shunosaurus

Spinosaurus

Elasmosaurus

Brachiosaurus

Quetzalcoatlus

Meet the giants

Not all dinosaurs, pterosaurs, and plesiosaurs were giants, but some of them were **really** huge. These enormous creatures were some of the largest animals ever. But which was the biggest of them all?

Argentinosaurus

Giganotosaurus
gig-AN-oh-toe-SORE-rus

Watch out, herbivores – there's a **giant** on the prowl! And this giant predator might have been even bigger than the legendary Tyrannosaurus!

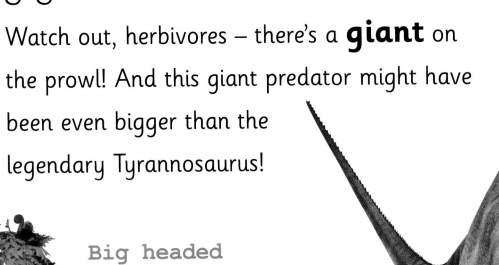

Big headed

With a powerful body and a skull the size of a person, Giganotosaurus was one of the **biggest** predators to have ever lived. Its name means "giant southern reptile", because it was found in South America.

Huge skull

251 million years ago
200
145
66

Triassic   Jurassic   Cretaceous

Fact File

Experts aren't exactly sure how big we were because not a lot of our fossils have been found.

Giganotosaurus had a thick, strong neck to support its huge head.

A giant meal

The monstrous Giganotosaurus was so big and strong it could take on **huge prey** – even enormous sauropods such as Argentinosaurus. Giganotosaurus may have worked as a team to bring down these massive plant-eaters.

Argentinosaurus

Size: 12m (39ft) long   **Habitat:** Woodlands   **Diet:** Meat

Diplodocus
dip-LOD-oh-kus

Spanning about the length of two school buses, Diplodocus was one of the **longest land animals** to have ever existed.

Extremely long tail

Not so gentle giants
Predators would have to think twice before attacking Diplodocus. Its huge size and **whip-like tail** made it very dangerous to attack.

Dippy
A cast of a diplodocus skeleton was on display for more than 100 years at the Natural History Museum in London. Its name was "**Dippy**".

251 million years ago | 200 | 145 | 66

Triassic | Jurassic | Cretaceous

Fact File

Diplodocus might have used its long neck to reach treetops. But some scientists think it may not have been able to lift its head very far, and only ate low-growing plants.

Small head

Long neck

Stumpy front legs

Diplodocus' tail was as long as the rest of its body.

Size: 25m (82ft) long   **Habitat:** Plains   **Diet:** Leaves

103

Shunosaurus

SHOE-noe-SORE-us

This **powerful sauropod** was very good at defending itself from attackers. It was first discovered in China in 1977.

Don't get too close or I'll **WHACK** you with my tail!

Odd one out

Like all sauropods, Shunosaurus had a long neck and tail, but its **neck was shorter** than other sauropods. This means it might have eaten from shorter trees than its relatives.

251 million years ago   200   145   66

Triassic   Jurassic   Cretaceous

Fact File

Wicked weapon

Shunosaurus is one of just a few known sauropods to have a **club on its tail**. This spiky club was most likely used to bash any dangerous predators – ouch!

Clubbed tail

Size: 10m (33ft) long   **Habitat:** Plains   **Diet:** Plants

Brachiosaurus

brackee-oh-SORE-uss

This huge sauropod had **longer front legs** than hind legs, so its body was sloped like a giraffe's. And just like a giraffe, Brachiosaurus got a lot of food from tall trees.

Small head compared to body

Head crest

Big bodies
Like all sauropods, Brachiosaurus had **powerful legs** to support their incredible weight.

We lived in what are now the forests of **NORTH AMERICA.**

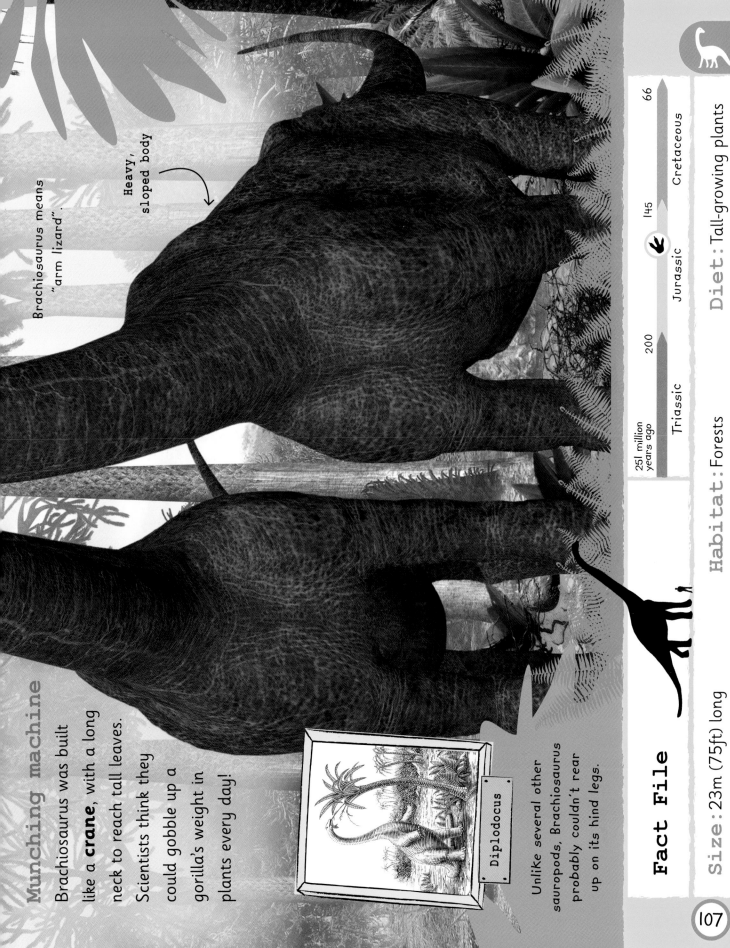

Munching machine

Brachiosaurus was built like a **crane**, with a long neck to reach tall leaves. Scientists think they could gobble up a gorilla's weight in plants every day!

Brachiosaurus means "arm lizard".

Heavy, sloped body

Unlike several other sauropods, Brachiosaurus probably couldn't rear up on its hind legs.

Diplodocus

Fact File

251 million years ago 200 145 66

Triassic Jurassic Cretaceous

Size : 23m (75ft) long

Habitat : Forests

Diet : Tall-growing plants

Quetzalcoatlus
Ket-zal-KWAT-luss

Quetzalcoatlus wasn't a dinosaur or a bird – it was a pterosaur. It was also **huge**, and possibly the biggest flying animal of all time.

> We could **FLY** at a speed of around 90kph (56mph).

> That thing is huge!

Big beak

Quetzalcoatlus' beak was 2.5m (8ft) long. That's longer than an adult person is tall! Its beak had no teeth so once it caught its prey, Quetzalcoatlus **swallowed** it whole.

251 million years ago	200	145		66
Triassic	Jurassic	Cretaceous		

Fact File

An enormous wingspan helped Quetzalcoatlus soar through the sky.

Air and land

Once Quetzalcoatlus spotted its prey from the air, it would land, fold up its wings, and quickly **chase after** it on the ground.

When we stood up we were as tall as giraffes.

As **FAST** as a lion

The **SIZE** of a small plane

With a beak as **LONG** as a person

Quetzalcoatl

Quetzalcoatlus was named after the Aztec god Quetzalcoatl, which was a giant flying snake.

Size: 11m (36ft) wingspan   **Habitat:** Plains   **Diet:** Small dinosaurs

Spinosaurus

SPINE-oh-SORE-us

Tyrannosaurus may get the glory, but Spinosaurus was actually the **biggest predator** to ever walk the Earth.

Fishing dinosaur

Not only did Spinosaurus have long, narrow jaws with razor-sharp teeth, but experts think little holes at the end of its snout helped it **sense fish** swimming around.

251 million years ago	200	145		66
Triassic		Jurassic		Cretaceous

Fact File

Sharp teeth

Wow. I can't believe how big Spinosaurus was!

Move over, Tyrannosaurus, I'm the **BIGGEST PREDATOR** around.

Only a few Spinosaurus bones have ever been discovered, so experts are still learning about it.

Spine lizard

Spinosaurus was named "spine lizard" for the **huge sail** on its back. It's not clear exactly what the sail was for, but some of the bones that made it up were bigger than a person!

Spinosaurus hunted giant fish, such as coelacanths.

Size: 16m (52ft) long   **Habitat:** Swamps   **Diet:** Fish

Argentinosaurus

ARE-jen-teen-oh-SORE-us

Possibly the **largest** and **heaviest** creature to ever walk on land, this giant was so big it made most other dinosaurs seem small!

Argentinosaurus egg

Big babies

An Argentinosaurus egg was about the size of a football. Scientists think it would have taken about **40 years** for a baby Argentinosaurus to grow to full size.

251 million years ago   200   145   66

Triassic   Jurassic   Cretaceous

Fact File

I weighed about the same as 20 **ELEPHANTS!**

Only a few Argentinosaurus bones have ever been found, so it's still a bit of a mystery, even to experts.

Our fossils were found in Argentina.

Feeding machines

Argentinosaurus needed to eat a lot to feed its massive body. Luckily, it didn't have much competition for food because its long neck helped it eat **tall leaves** that other dinosaurs couldn't reach.

Long neck

Size: 35m (115ft) long   **Habitat:** Forests   **Diet:** Plants

Elasmosaurus
el-LAZZ-moe-SORE-us

This bizarre sea monster lived at the same time as the dinosaurs, but was actually a **plesiosaur**. Its enormous neck was longer than the rest of its body!

Neck or tail?
Scientists had never seen anything like Elasmosaurus when its fossils were first discovered. Its neck was **so long** they thought it was a tail and drew sketches with the head at the wrong end of the body.

251 million years ago

200

145

66

Triassic

Jurassic

Cretaceous

Fact File

Some scientists think Elasmosaurus could coil its neck up like a snake, but others think it wasn't very flexible.

Mythical monsters

Some people think the Scottish Loch Ness Monster and other mythical **sea monsters** from around the world are really Elasmosaurus. But no, Elasmosaurus went extinct 66 million years ago, at the same time as the dinosaurs.

Elasmosaurus was probably quite a slow swimmer.

Elasmosaurus hunted near the seabed and came up to breathe.

My neck had 72 bones in it. That's more than any other creature.

Sharp teeth for catching prey

Size: 14m (45ft) long   **Habitat:** Oceans   **Diet:** Fish and squid

Anchiornis

Sinosauropteryx

Compsognathus

Microraptor

Meet the mini monsters

Don't forget about the little guys! These **dinky dinosaurs** were tiny, but some of them were also clever, crafty, and fierce. Find out how these fun-sized creatures more than made up for being so mini.

Caudipteryx

Caudipteryx
kor-DIP-ter-iks

With its feathered body and tail, this little dinosaur might have looked a bit like a prehistoric **peacock**.

Adaptable eater

Caudipteryx was a theropod. Most theropods only ate meat but Caudipteryx **ate plants and animals**. Its toothy beak could shred plants, gobble insects, and crack open seeds.

We had feathers but couldn't fly. Our feathers kept us warm and helped us look good to other Caudipteryx.

251 million years ago	200	145	66
Triassic	Jurassic	Cretaceous	

Fact File

One of my most interesting features was the feathers at the end of my tail. In fact, my name means "tail feathers".

Common dinosaur

Quite a lot of Caudipteryx fossils have been found in the same place in China. This might mean that lots of these little dinosaurs **lived together** in groups.

Size: 1m (3ft) long   **Habitat:** Rivers   **Diet:** Plants and small animals

Compsognathus
COMP-sog-NAITH-us

Who says you need to be **big** to be a predator?
Compsognathus was small, but had pointy teeth,
sharp claws, and swift legs to help it hunt.

We were good hunters,
but if we found a dead
animal we weren't above
scavenging either.

Mini but menacing
Compsognathus was only a little **taller than a chicken**, but what it lacked in size it made up for in speed. Its hollow bones helped it stay light on its feet and it ran on tiptoes to be as quick as possible.

251 million
years ago   200   145   66

Triassic   Jurassic   Cretaceous

Fact File

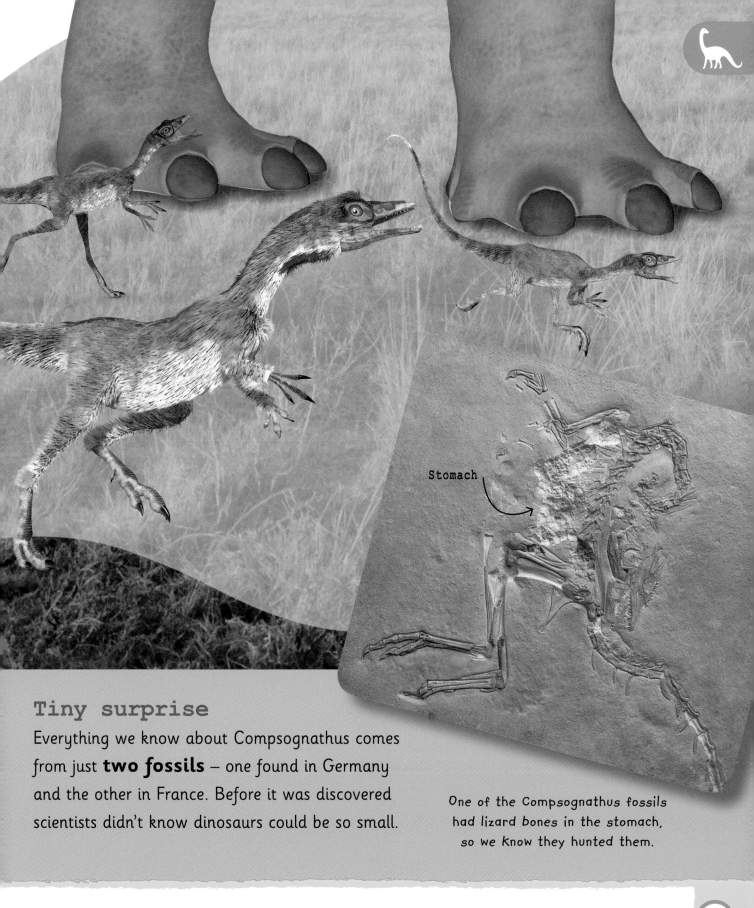

Stomach

Tiny surprise

Everything we know about Compsognathus comes from just **two fossils** – one found in Germany and the other in France. Before it was discovered scientists didn't know dinosaurs could be so small.

One of the Compsognathus fossils had lizard bones in the stomach, so we know they hunted them.

Size: 1m (3ft) long   **Habitat:** Scrublands   **Diet:** Small animals

Eoraptor

EE-oh-RAP-tor

One of the **earliest dinosaurs**, Eoraptor was about the same size as a small dog. Unlike most friendly pets – it was a **fierce** hunter.

My blade-like teeth would have easily sliced through meat, but some scientists think I might have eaten plants, too.

Eoraptor could see well in almost all directions, which would have been great for spotting prey.

Dawn of the dinosaurs
The name Eoraptor means **"dawn thief"** because it was alive at the beginning, or dawn, of the age of the dinosaurs.

Sharp claws

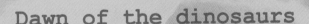

251 million years ago	200	145	66
Triassic	Jurassic	Cretaceous	

Fact File

Scientists aren't sure yet whether Eoraptor had **FEATHERS** or **SCALES**.

Claw

Two clawless, smaller fingers

Give me five!

As an early dinosaur, some of Eoraptor's features were not very well developed. It had five fingers at the end of its front limbs, but only three of these fingers **had claws**.

Long, speedy legs

Size: 1m (3ft) long　　**Habitat:** River valleys　　**Diet:** Small animals

Sinosauropteryx

SIGH-no-SORE-op-ter-ix

A Sinosauropteryx fossil found in China in 1996 turned out to be one of the most important dinosaur discoveries. Why? It showed they could have **feathers**!

Fantastic feathers

Until Sinosauropteryx was discovered everyone thought all dinosaurs were **scaly**. But the Sinosauropteryx fossil had clear, fuzzy marks around its body that scientists realized were **feathers** to keep it warm.

Many other incredible fossil discoveries have been made in the Liaoning Province in China.

This Sinosauropteryx fossil **CHANGED THE WAY** experts

251 million years ago 200 145 66

Triassic Jurassic Cretaceous

 Fact File

Sinosauropteryx means "Chinese reptilian wing".

Outline of feathers

Grounded

Sinosauropteryx's feathers were soft and downy, like an ostrich's. Just like an ostrich, Sinosauropteryx **couldn't fly** but had long, strong legs and could run very fast.

thought about dinosaurs forever.

Size: 1m (3ft) long   **Habitat:** Forests   **Diet:** Small animals

125

Anchiornis

AN-kye-OR-niss

One of the **smallest** dinosaurs ever, the feathered Anchiornis was only about the size of a magpie.

> We probably couldn't fly properly, but we could glide down to catch little bugs.

Almost a bird

Scientists named this little dinosaur Anchiornis, which means **"near bird"** because it was covered from head to toe in feathers.

Anchiornis fossil

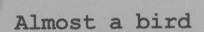

Fact File

Colourful head

Scientists have studied fossils of Anchiornis' feathers and think it was **black and grey**, with a tuft of **red feathers** on its head.

Scientists used a very special microscope to try to work out what colour I was.

Size: 40cm (16in) long   **Habitat:** Woodlands   **Diet:** Mostly insects

Microraptor

MY-crow-rap-tor

This **tiny** bird-like dinosaur was a vicious hunter from the Cretaceous. Its body was about the same size as a rabbit, and covered in feathers.

The long feathers on my arms and legs are excellent for gliding, but they made me clumsy while running.

Many well-preserved Microraptor fossils have been discovered in China.

251 million years ago

200

145

66

Triassic

Jurassic

Cretaceous

Fact File

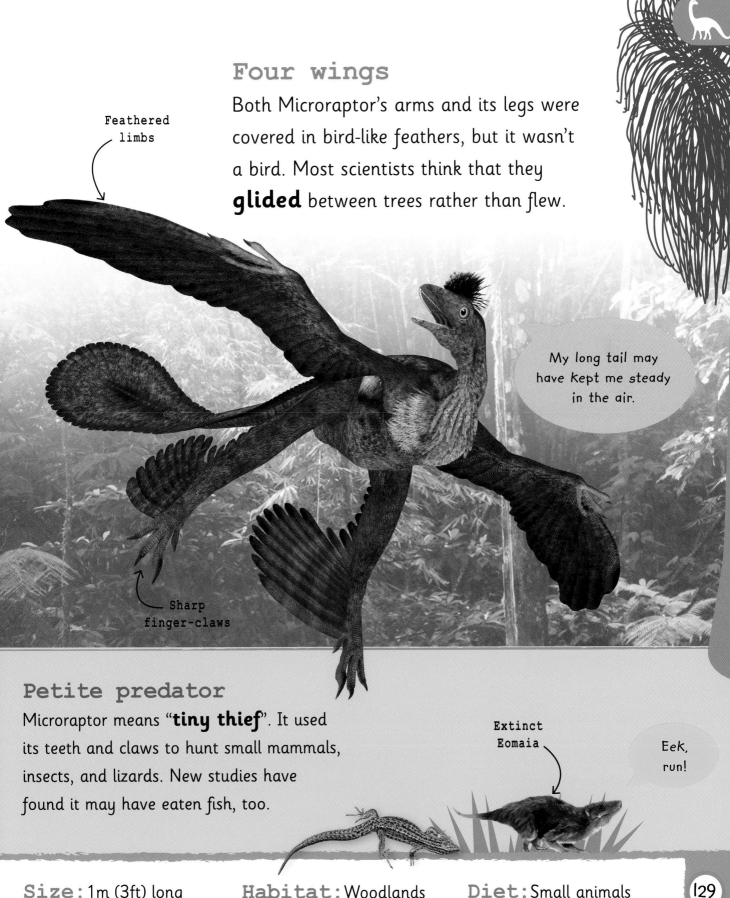

Four wings

Both Microraptor's arms and its legs were covered in bird-like feathers, but it wasn't a bird. Most scientists think that they **glided** between trees rather than flew.

Feathered limbs

My long tail may have kept me steady in the air.

Sharp finger-claws

Petite predator

Microraptor means "**tiny thief**". It used its teeth and claws to hunt small mammals, insects, and lizards. New studies have found it may have eaten fish, too.

Extinct Eomaia

Eek, run!

Size: 1m (3ft) long **Habitat:** Woodlands **Diet:** Small animals

Pachycephalosaurus

Amargasaurus

Sauropelta

Stegosaurus

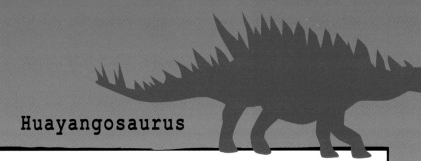

Huayangosaurus

Meet the sharp, bumpy, and spiky ones

These dinosaurs went up against some of the fiercest predators to have ever lived. They stuck up for themselves with **strong armour** and powerful defensive weapons. Meet the herbivores who fought back!

Euoplocephalus
YOU-owe-plo-SEFF-ah-luss

Many dinosaurs **defended** themselves from attackers using their **teeth** or **claws**, but Euoplocephalus had another way — it was covered in protective armour.

Terrible tail

If attacked, Euoplocephalus had a huge, heavy club at the end of its tail that it could swing at attackers with **bone-crushing power.**

Euoplocephalus's armoured body was built like a tank.

251 million years ago   200   145   66

Triassic   Jurassic   Cretaceous

Fact File

A walking tank

Euoplocephalus was twice the size of a rhinoceros and covered head to toe in **armoured** plates made from bone — even its eyelids had armour.

Even though I was heavy, experts think I was probably quite fast for my size.

Spiky plates

A soft belly was its only weak spot.

Size: 6m (20ft) long **Habitat:** Woodlands **Diet:** Plants

133

Triceratops
try-SERRA-tops

With its impressive horns and frill, Triceratops had some serious **defensive weapons**. And with Tyrannosaurus on the prowl, they needed them!

Frill

Horn

Thick head

Triceratops would fight over mates and bump their heads together, so they needed **strong heads** to survive these fights. These hard skulls made high-quality fossils.

251 million years ago · 200 · 145 · 66

Triassic · Jurassic · Cretaceous

Fact File

Defensive weapons

Fossils have shown that Tyrannosaurus and Triceratops often got into nasty fights. And as powerful as Tyrannosaurus was, Triceratops was not **easy prey**. Its neck frill acted as a thick shield and its horns were sharp, long, and strong.

We were about the size of an ELEPHANT but looked more like a RHINOCEROS.

Triceratops munched on low-growing ferns and probably lived in herds for safety.

Size: 9m (30ft) long   **Habitat:** Forests   **Diet:** Plants

Amargasaurus

ah-MAR-gah-SORE-us

Although it was quite small compared to other sauropods, Amargasaurus stood out from the crowd because of its striking and unusual **neck**.

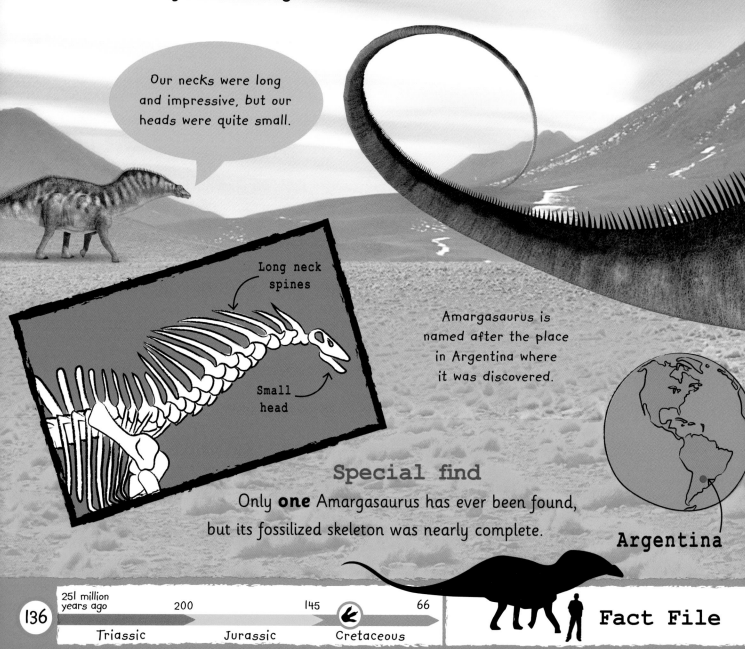

Our necks were long and impressive, but our heads were quite small.

Long neck spines

Small head

Amargasaurus is named after the place in Argentina where it was discovered.

Argentina

Special find

Only **one** Amargasaurus has ever been found, but its fossilized skeleton was nearly complete.

251 million years ago	200	145	66
Triassic	Jurassic	Cretaceous	

Fact File

Spikes or sail?

The double rows of **long bones** sticking out of Amargasaurus's neck and back might have been spikes or a sail. They could have been used to impress other dinosaurs or even to shake and rattle to make noise.

We had the tallest spikes of any sauropod.

Size: 11m (36ft) long   **Habitat:** Scrublands   **Diet:** Plants

Stegosaurus
STEG-oh-SORE-uss

With two rows of huge **bony plates** along its back, it's easy to see why Stegosaurus is called "plated lizard".

Big dino, little brain

Stegosaurus had a tiny brain, even for a dinosaur. Stegosaurus was about the size of an elephant, but its brain was no bigger than an **apple**. It probably wasn't the smartest of dinosaurs!

251 million years ago | 200 | 145 | 66

Triassic | Jurassic | Cretaceous

Fact File

Bony plates

Show offs

Although Stegosaurus' plates look impressive, experts think they were just for show. They were possibly used to attract a mate or warn off predators.

Allosaurus fossils have been found with holes made by Stegosaurus tails.

Spiked tail

Stegosaurus had a sharp toothless beak to nibble on leaves and ferns.

Our dangerous spiked tails came in handy if we were ever under attack.

Size: 9m (30ft) long   **Habitat:** Woodland   **Diet:** Plants

Styracosaurus

sty-RACK-oh-SORE-us

While many dinosaurs had spikes, horns, or frills, Styracosaurus had **all three** – making it a seriously impressive looking herbivore!

The holes in Styracosaurus' skull were covered by skin.

Nose horn

The name Styracosaurus means "spiked lizard".

Skull

Horny head

Although its frill spikes looked impressive, they weren't very useful for fighting, and were just for show. Styracosaurus probably fought with its **nose horn** instead.

251 million years ago   200   145   66

Triassic   Jurassic   Cretaceous

Fact File

Our enormous skulls were about 1.8m (6ft) long – longer than an average adult person.

Frill spikes

Fully featured

Styracosaurus was smaller than its famous relative **Triceratops**, but it had a few extra features. To go with its frill, it had a nose horn, cheek horns, and spikes running along its back and tail.

Size: 5m (17ft) long   **Habitat:** Woodlands   **Diet:** Plants

Pachycephalosaurus
PACK-ee-sef-ah-low-SORE-us

There is probably a very good reason that Pachycephalosaurus had such a **big, thick skull**, but scientists still aren't certain what it is.

Our skulls were **20 TIMES THICKER** than other dinosaur skulls.

251 million years ago   200   145   66

Triassic   Jurassic   Cretaceous

 **Fact File**

Helmet head

Pachycephalosaurus' special bony head was so hard it was probably for **protection**, but what exactly from?

1 One theory is Pachycephalosaurus fought each other, banging heads to **show dominance** like stags. This might explain why they needed a thick skull — but there is no evidence to prove it.

Ouchh!

2 It's also possible that Pachycephalosaurus used its head like a **battering ram** to fight off other dinosaurs that got too close.

BASH!

Size: 5m (16ft)   **Habitat:** Forests   **Diet:** Plants, fruit, and seeds

Sauropelta
SORE-oh-PELT-ah

The sturdy Sauropelta was about the same size as a **rhinoceros**, but because of its thick, studded armour, it would have been a lot heavier.

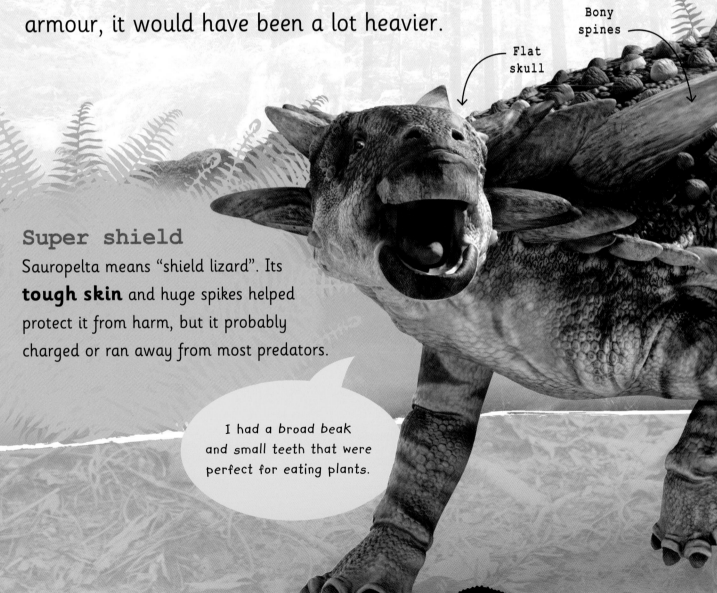

Flat skull

Bony spines

Super shield

Sauropelta means "shield lizard". Its **tough skin** and huge spikes helped protect it from harm, but it probably charged or ran away from most predators.

I had a broad beak and small teeth that were perfect for eating plants.

251 million years ago · 200 · 145 · 66

Triassic · Jurassic · Cretaceous

Fact File

Predators beware!

Sauropelta had big **spines** that stuck out from its neck, and a thick layer of bony armour that looked like a knight's chain mail. These tools would have made life tough for meat-eating dinosaurs looking for a meal.

Studded armour

Short legs

Size: 5m (16ft) long   **Habitat:** Woodlands   **Diet:** Plants

Huayangosaurus
hoy-YANG-oh-SORE-uss

This herbivore was one of the earliest members of the **stegosaur** family. Although it was smaller than its famous relative Stegosaurus, the family resemblance is easy to spot.

My spiky tail helped **PROTECT** me from predators.

251 million years ago   200   145   66

Triassic   Jurassic   Cretaceous

Fact File

Fossil finds

Huayangosaurus fossils were discovered with lots of different dinosaurs in a quarry that used to be a giant lake in **China**. More than 8,000 fossils have been found there.

Dinosaur family

They lived in different places and millions of years apart, but Huayangosaurus and Stegosaurus looked similar. They both had **impressive back plates** and **tail spikes**, but Huayangosaurus' plates were more pointy and it also had shoulder spikes.

Stegosaurus

Huayangosaurus

More pointed back plates

All legs the same length

Longer back legs

Size: 4m (13ft) long   **Habitat:** River valleys   **Diet:** Plants

147

Lambeosaurus

Therizinosaurus

Gallimimus

Ouranosaurus

Tupandactylus

Meet
the show-offs

With their crests, colours, feathers, and sails, these weird and wonderful reptiles really **stood out from the crowd**. Discover why being the biggest show-off around could be very useful in the Mesozoic Era.

Guanlong

Lambeosaurus
LAM-bee-oh-SORE-uss

This big plant-munching dinosaur is most famous for its unusual head crest. In fact, Lambeosaurus is the only dinosaur discovered so far with **two prongs** on its crest.

Bony crest

Two pronged crest

Rows of teeth

We belonged to a group of dinosaurs called hadrosaurids.

Mystery crest
Lambeosaurus' crest has puzzled scientists since it was discovered. No one knows what it was used for, but it was probably used to **show off** to other Lambeosauruses.

251 million years ago   200   145   66

Triassic   Jurassic   Cretaceous

Fact File

Toothy herbivore

Lambeosaurus had a bill similar to a duck's, but it also had **hundreds of teeth** to mash up leaves. Its teeth regrew throughout its life to make sure they were always healthy.

We were named after the scientist Lawrence Lambe, who discovered the first Lambeosaurus fossil.

Size: 9m (30ft) long   **Habitat:** Woodlands   **Diet:** Plants and leaves

Ouranosaurus
oo-RAH-no-SORE-uss

Ouranosaurus was a large herbivore that lived near rivers in what is now **Africa**. It spent a lot of its time munching away on plants.

Experts think that I moved very slowly.

251 million years ago

200

145

66

Triassic

Jurassic

Cretaceous

Fact File

Sailing ahead

Ouranosaurus had an impressive bony sail on its back that was covered in scaly skin. Scientists think that Ouranosaurus used its sail to **attract mates** or possibly help it cool down in hot weather.

Sail

Ouranosaurus' sail was similar to Spinosaurus', but they were very different dinosaurs.

Ouranosaurus was named after the Ancient Greek god **OURANOS**, but its name also means "brave lizard".

Niger

Secret in the sand

Ouranosaurus fossils were **found in a desert** in Niger. The fossils were easy to dig out by hand because they were only loosely buried in the sand.

Size: 7m (23ft) long   **Habitat:** Riverbanks   **Diet:** Plants

Tsintaosaurus

SIN-tow-SORE-uss

This strange-looking dinosaur had a crest on top of its head that has **confused** scientists since it was discovered.

> I belonged to a group of dinosaurs called hadrosaurs. We had beaks like a duck's.

Front or back?

Scientists first thought Tsintaosaurus' crest pointed **forward** out of its head like a unicorn's horn. But they had only found a part of it. We now know it actually pointed **backwards**.

This herbivore walked on four legs but could stand on two.

251 million years ago   200   145   66

Triassic   Jurassic   Cretaceous

Fact File

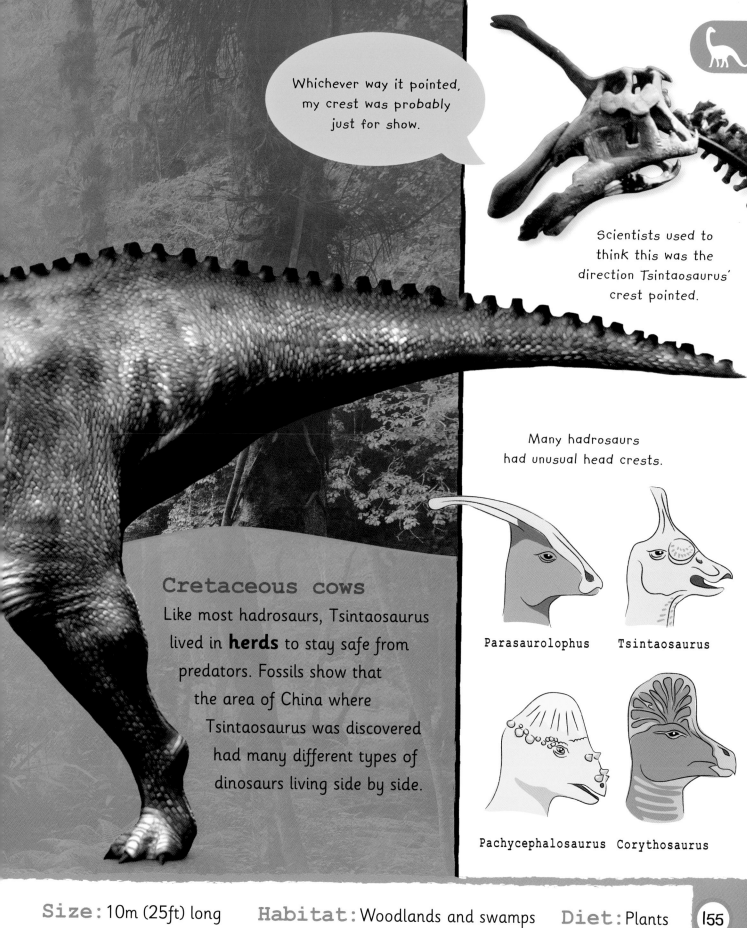

Whichever way it pointed, my crest was probably just for show.

Scientists used to think this was the direction Tsintaosaurus' crest pointed.

Many hadrosaurs had unusual head crests.

Parasaurolophus   Tsintaosaurus

Pachycephalosaurus   Corythosaurus

Cretaceous cows

Like most hadrosaurs, Tsintaosaurus lived in **herds** to stay safe from predators. Fossils show that the area of China where Tsintaosaurus was discovered had many different types of dinosaurs living side by side.

Size: 10m (25ft) long   Habitat: Woodlands and swamps   Diet: Plants   155

Guanlong
GWON-long

As the earliest known dinosaur belonging to the tyrannosaur group, Guanlong was an early relative of **T. rex**, but they lived 90 million years apart.

Crest or crown?
Guanlong fossils were first discovered in China in 2006. Its name comes from the Chinese name for **"crown dragon"**, because of the bony crest on top of its head.

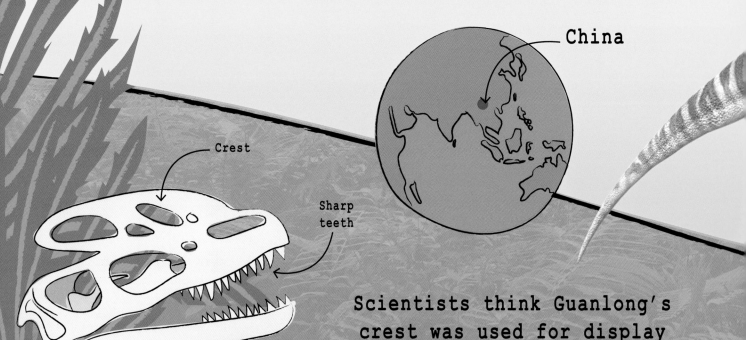

China

Crest

Sharp teeth

Scientists think Guanlong's crest was used for display and attracting a mate.

251 million years ago | 200 | 145 | 66

Triassic | Jurassic | Cretaceous

Fact File

Most dinosaurs in the tyrannosaur family had **TWO** fingers, but we had **THREE**.

It's likely that Guanlong's body was covered in fur or feathers.

Fairly short arms

Guanlong hunted small dinosaurs and reptiles.

Size: 3m (10ft) long   **Habitat:** Woodlands   **Diet:** Meat

Corythosaurus

Ko-RITH-oh-SORE-us

When scientists discovered Corythosaurus, they thought that the **crest** on its head looked like helmets worn by Ancient Greek soldiers, so they named it "helmet lizard".

Although Corythosaurus walked on all fours, it may have also been able to walk on two legs.

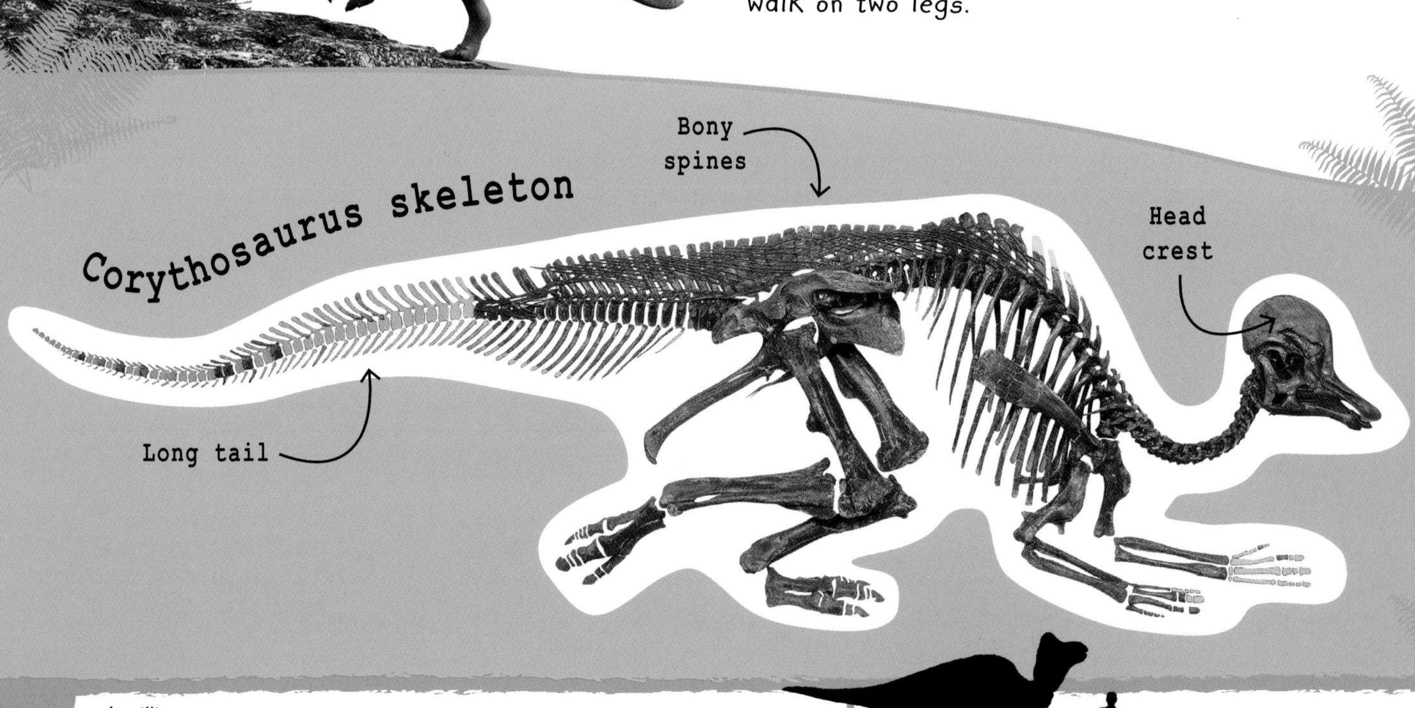

Corythosaurus skeleton

Bony spines

Head crest

Long tail

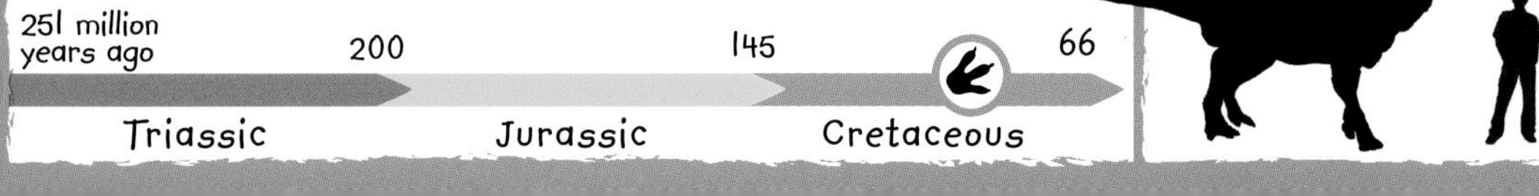

251 million years ago 200 145 66

Triassic Jurassic Cretaceous

Fact File

We had a row of bony **SPINES** all the way along our backs.

It's likely that Corythosaurus ate fruit and pine needles as well as twigs, leaves, and seeds.

Musical marvel?

Corythosaurus' crest was hollow, and connected to its nose. This meant it may have worked a little like a **trumpet**, and could have been used to communicate with others.

Size: 9m (30ft) long   **Habitat:** Forests   **Diet:** Leaves and seeds

Therizinosaurus

THERRY-zin-oh-SORE-us

With its long neck, pot belly, huge claws, and stumpy feet, Therizinosaurus was one of the **strangest** dinosaurs around.

Treetop dining

Its long neck and tall frame would have helped Therizinosaurus reach the branches of **tall trees** to gather leaves, just like giraffes.

Feathered body

Enormous claws

Crazy claws

Fossils show that Therizinosaurus had claws the size of **baseball bats**. As a plant-eater, it probably only used these claws to defend itself or to pull down branches.

Strong legs

Odd one out

Therizinosaurus belonged to a group of dinosaurs called **theropods**, who, unlike Therizinosaurus, were mostly meat-eating predators.

We had large pot bellies because we needed big digestive systems to handle all the plants we ate.

Fact File

Size: 8m (26ft) long

Habitat: Forests

Diet: Plants

251 million years ago

200 | 145 | 66

Triassic | Jurassic | Cretaceous

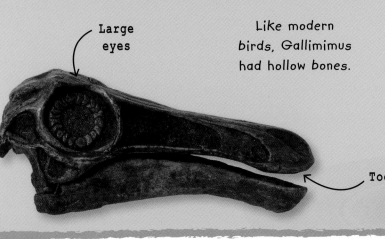

My strong legs and light body made me the perfect sprinter.

Gallimimus
GAL-ih-MIME-us

If there's one thing Gallimimus is famous for, it's being **super speedy**. It was probably the fastest dinosaur of all time.

Large eyes

Like modern birds, Gallimimus had hollow bones.

Toothless beak

Stone swallower

Gallimimus didn't have any teeth, so it ate stones, which helped digest leaves by **grinding them up** inside its stomach.

251 million years ago
200
145
66

Triassic
Jurassic
Cretaceous

Fact File

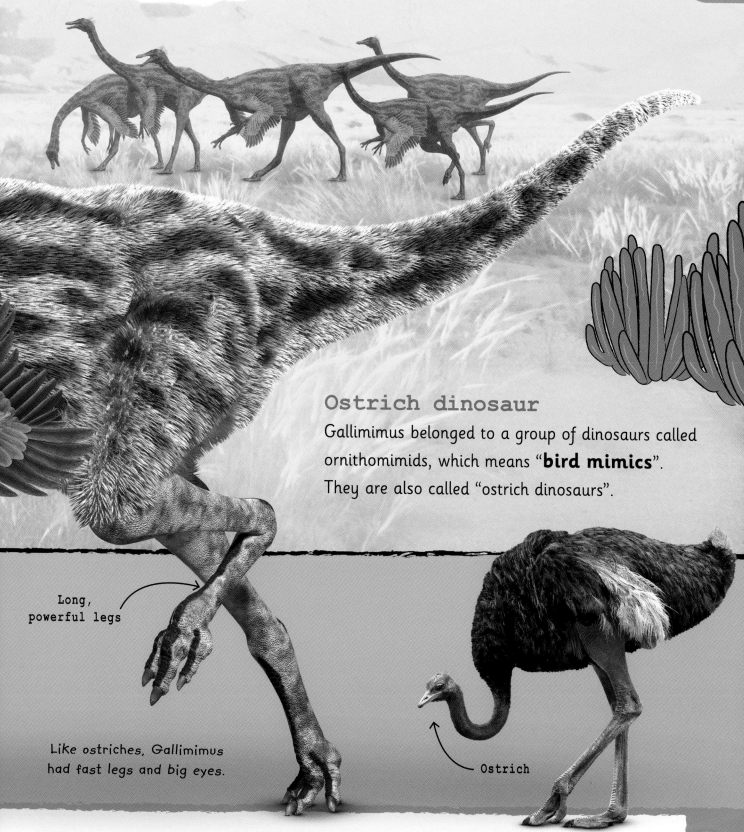

Ostrich dinosaur

Gallimimus belonged to a group of dinosaurs called ornithomimids, which means "**bird mimics**". They are also called "ostrich dinosaurs".

Long, powerful legs

Like ostriches, Gallimimus had fast legs and big eyes.

Ostrich

Size: 6m (20ft) long   **Habitat:** Desert plains   **Diet:** Leaves and insects

Tupandactylus
too-pan-DAK-til-us

This winged wonder had an **elaborate crest** on its head. This crest was probably very colourful, and was used to **show off** and attract mates.

Soaring in the sky

Is it a bird? Is it a bat? No, it's a **pterosaur**! Pterosaurs had wings made of bone, muscle, and a thin layer of skin, like bat wings. Unlike bats, which are mammals, pterosaurs were reptiles.

Whopping wingspan

At 3m (9ft), the **albatross** has the largest wingspan of any creature today. Tupandactylus' wingspan was almost twice as big!

Long wings

251 million years ago   200   145   66

Triassic   Jurassic   Cretaceous

Fact File

Crestfallen

Its crest was so huge that it might have made it hard for Tupandactylus to fly. It's possible Tupandactylus spent a lot of time **on land**, folding its wings and walking around.

Tupandactylus's crest was made of keratin, which is the same substance our fingernails are made from.

It might not be that practical, *but* isn't my crest impressive?

Size: 4m (13ft) wingspan   **Habitat:** Coasts   **Diet:** Fish and possibly fruit

Pteranodon

Dimorphodon

Hamipterus

Meet the sky soarers

Long before birds existed, these fantastic flying reptiles took to the skies on their leathery wings. They weren't dinosaurs, but they lived alongside them through most of the Mesozoic Era. Say hello to the **pterosaurs**.

Rhamphorhynchus

Archaeopteryx

ar-kee-OP-ter-ix

It may be **small**, but when Archaeopteryx was found it was one of the biggest, most important discoveries in the history of palaeontology.

Beak with teeth

The missing link

The discovery of Archaeopteryx proved there is a link between **dinosaurs** and **birds**. Archaeopteryx is one of the oldest members of the bird family!

Archaeopteryx had features of both reptiles and birds.

251 million years ago   200   145   66

Triassic   Jurassic   Cretaceous

 Fact File

Flying reptiles called pterosaurs had existed for a while, but I was probably the first dinosaur that could fly.

Feathered wings

Feathered tail

Germany

Fantastic feathers

An amazing detailed Archaeopteryx fossil found in Germany clearly shows markings of feathers on its **wings** and **tail**.

Size: 0.3m (1ft)   **Habitat:** Forests   **Diet:** Insects and small reptiles

Pteranodon

teh-RAN-oh-don

High in the sky 80 million years ago, flocks of Pteranodons with huge **wings** were gliding and swooping in the air. It would have been an amazing sight!

Built for speed

Pteranodon's kite-like body helped it fly at high speeds while looking for a meal. It **soared** on the wind, only flapping its wings occasionally.

We flew around in large flocks.

Pteranodons had LONG, THIN BEAKS perfect...

It's possible that Pteranodon's head crest was just for display.

251 million years ago	200	145	66
Triassic	Jurassic	Cretaceous	

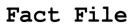

Fact File

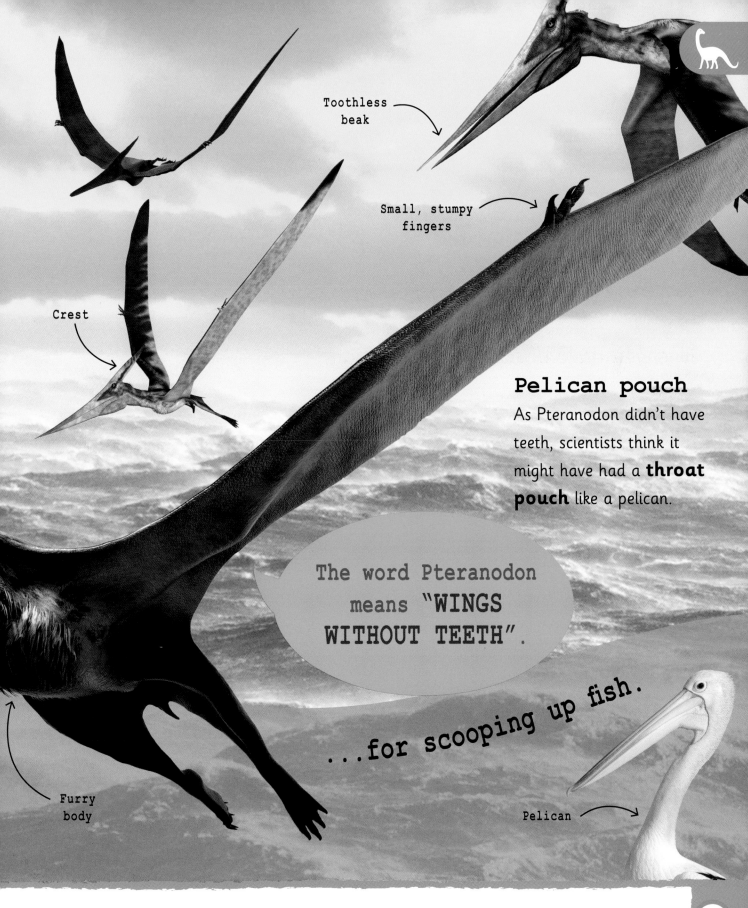

Toothless
beak

Small, stumpy
fingers

Crest

Furry
body

Pelican pouch

As Pteranodon didn't have teeth, scientists think it might have had a **throat pouch** like a pelican.

The word Pteranodon means "WINGS WITHOUT TEETH".

...for scooping up fish.

Pelican

Size: 9m (30ft) wingspan   **Habitat:** Coasts   **Diet:** Fish

Hamipterus
ham-IP-ter-us

In 2017, around 200 fossilized Hamipterus eggs were found in China. This incredible discovery helped scientists learn **more** about pterosaurs than ever before.

Finding this many eggs in one place means that flocks of Hamipterus probably nested as groups.

Bones

Eggs

Flightless babies

Some of the eggs had preserved babies inside that had undeveloped wing bones. Some scientists think this means newborns **couldn't fly** straight after hatching, but other scientists disagree.

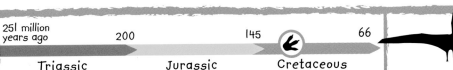

251 million years ago   200   145   66

Triassic   Jurassic   Cretaceous

Fact File

Large crest

Important discovery

Fossilized pterosaur eggs are **very rare**. The discovery in China is a real treasure because scientists can now study how pterosaur babies developed and grew.

Experts think pterosaurs grew inside eggs for a long time before hatching.

Folded wings

Our long teeth helped us catch fish to feed our young.

Size: 3m (10ft) wingspan　　**Habitat:** Rivers and lakes　　**Diet:** Fish

Rhamphorhynchus
ram-foe-RINK-us

A small but **powerful flyer**, Rhamphorhynchus soared above the Jurassic waters gobbling up all the fish that it could catch.

Diamond-shaped flap of skin

Coastal hunter

Rhamphorhynchus lived near coasts and rivers. It flew above the water then **swooped down** to catch fish, gripping them between its sharp, pointy teeth.

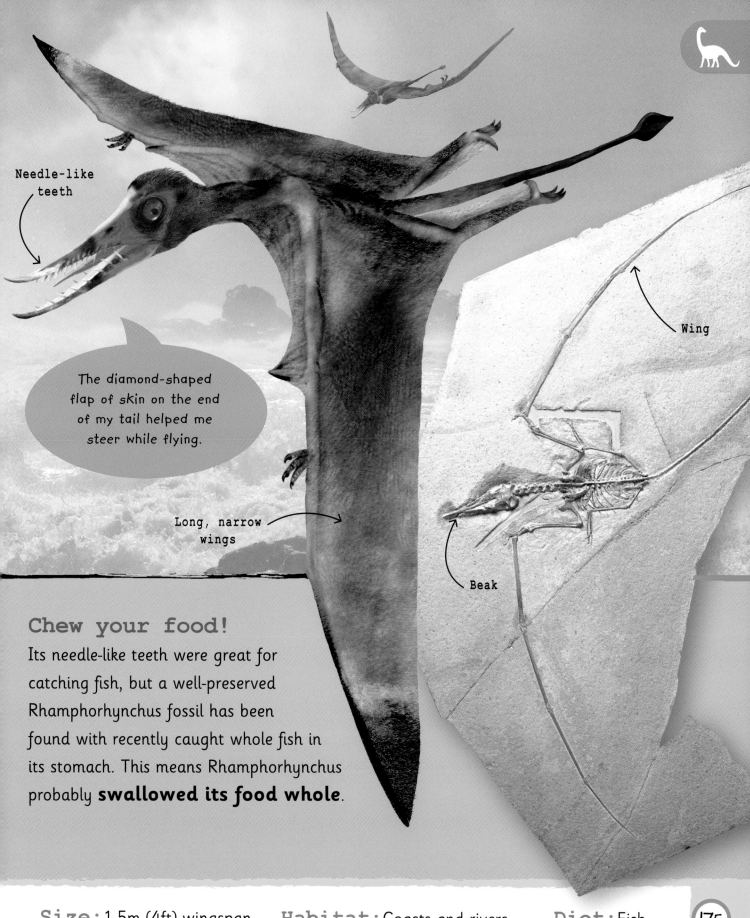

Needle-like teeth

The diamond-shaped flap of skin on the end of my tail helped me steer while flying.

Long, narrow wings

Wing

Beak

Chew your food!

Its needle-like teeth were great for catching fish, but a well-preserved Rhamphorhynchus fossil has been found with recently caught whole fish in its stomach. This means Rhamphorhynchus probably **swallowed its food whole**.

Size: 1.5m (4ft) wingspan   **Habitat:** Coasts and rivers   **Diet:** Fish

Dimorphodon
dye-MOR-foh-don

There's a reason this tiny pterosaur looked quite different from its relatives — it was one of the **first pterosaurs** around.

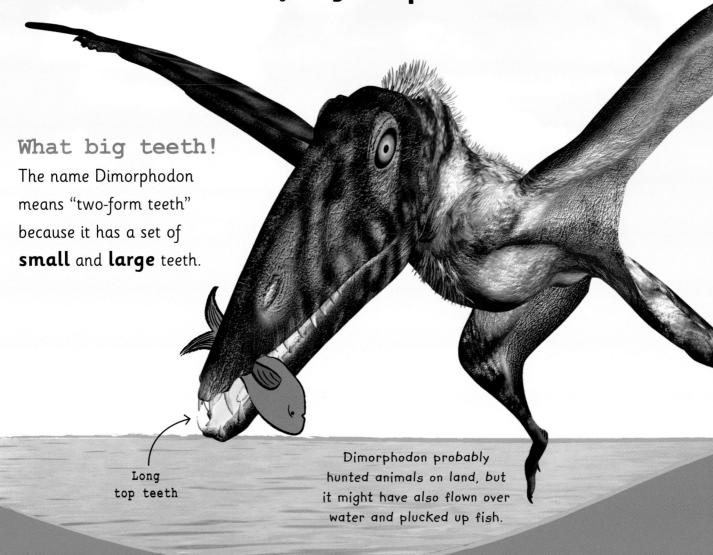

What big teeth!

The name Dimorphodon means "two-form teeth" because it has a set of **small** and **large** teeth.

Long
top teeth

Dimorphodon probably hunted animals on land, but it might have also flown over water and plucked up fish.

251 million
years ago   200   145   66

Triassic   Jurassic   Cretaceous

Fact File

Tree pterosaur

Unlike the pterosaurs that came after it, Dimorphodon wasn't a very good flyer and could only stay airborne for a little while. But it was a good **climber** and could scurry up trees like a squirrel.

Fossilized
Dimorphodon

I had a big head that was about a third of my *body* length.

Dimorphodon walked on four legs, but not very fast!

Mosasaurus

Rhomaleosaurus

Liopleurodon

Ichthyosaurus

Meet the creatures of the deep

While the dinosaurs ruled the land, the oceans of the Mesozoic were teeming with amazing **sea-dwelling reptiles**. Take a deep breath and dive into these pages to discover what lurked in the prehistoric waters.

Mosasaurus
MOSE-ah-saw-rus

One of the last giant marine reptiles, this **enormous** crocodile-like beast terrorized the waters of the Late Cretaceous.

Skull the size of a person

Modern Mosasaurus

Scientists think that this enormous reptile is a distant relative of modern land animals such as snakes and the **monitor lizard**.

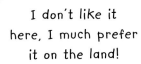

I don't like it here, I much prefer it on the land!

Monitor lizard

251 million years ago	200	145	66
Triassic	Jurassic	Cretaceous	

Fact File

Our **FLEXIBLE** jaws allowed us to swallow smaller prey **WHOLE**.

A long, flat tail helped Mosasaurus propel itself through water.

← Large flippers

What's for dinner?

Mosasaurus ate **anything** from squid and fish to turtles, ammonites, and possibly even small dinosaurs that got too close to the water.

Size: 15m (50ft) long   **Habitat:** Oceans   **Diet:** Fish and marine reptiles

Liopleurodon
LIE-oh-PLOOR-oh-don

This hefty beast was a surprisingly **speedy swimmer**. It was built like a whale, and was able to glide through water with ease.

Flaptastic flippers

The secret to Liopleurodon's swift swimming was its **flippers**. It beat its front and back flippers up and down as pairs, pushing Liopleurodon quickly through the water.

Powerful flippers

251 million years ago	200		145	66
Triassic		Jurassic	Cretaceous	

Fact File

Mighty bite

The deadliest prehistoric predators didn't only live on land. Liopleurodon's jaws were **just as strong** as those of most dinosaurs, including the mighty Tyrannosaurus.

Monstrous teeth

Like all sea-dwelling reptiles, I came to the surface to breathe.

Super smell

It wasn't just speed and power that made Liopleurodon such a deadly ocean predator. Experts think it had an excellent **sense of smell** that was great for hunting prey.

Nostril

Size: 7m (23ft) long   **Habitat:** Seas   **Diet:** Marine animals

Ichthyosaurus
ICK-thee-oh-SORE-uss

Although it looks a lot like a dolphin, Ichthyosaurus was actually a speedy **sea-dwelling reptile** from the Early Jurassic.

Strong swimmers

Just like a dolphin, Ichthyosaurus was an excellent ocean hunter. But while dolphins have great hearing, Ichthyosaurus used its **sharp eyesight** to track prey in the dark water.

Pointy teeth

Huge eye

251 million years ago | 200 | 145 | 66

Triassic | Jurassic | Cretaceous

 Fact File

Our long jaws and thin, pointed teeth made catching fish easy work.

Smooth skin

Flexible fin

Swimming start

Unlike most reptiles, Ichthyosaurus gave birth to live young instead of laying eggs. The babies were born **tail first** so that they didn't drown.

Long, thin jaw

Size: 2m (6½ft) long   **Habitat:** Oceans   **Diet:** Fish and squid

Rhomaleosaurus
ROME-alley-oh-SORE-us

The master of the Early Jurassic seas, this powerful **plesiosaur** lurked in the depths terrorizing fish, squid, and other marine reptiles that got in its way.

We had pointed teeth like a crocodile's. They were perfect for clutching slippery prey.

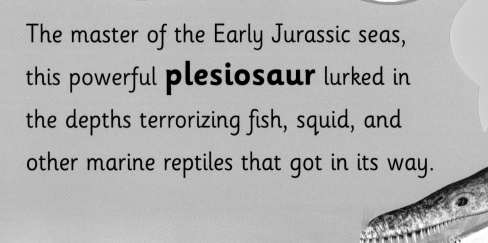

Pointed teeth

We flapped our flippers like wings to "fly" through the water at great speed.

251 million years ago

200

145

66

Triassic

Jurassic

Cretaceous

Fact File

Hunter's tools

A fast swimmer with **keen eyesight** and a great sense
of smell, Rhomaleosaurus tracked its prey then charged
toward it by beating its huge flipper-like fins.

Short
neck

Fins

Rhomaleosaurus had quite a
short neck for a plesiosaur.

Strong lizard

Rhomaleosaurus was
such a fierce predator that
its name means "strong
lizard". Once it caught
prey, it may have **spun
around** to rip it into
smaller pieces to swallow.

Size: 7m (21ft) long   **Habitat:** Oceans   **Diet:** Fish and squid

Kronosaurus
crow-no-SORE-us

I used my huge **FLIPPERS** to soar through the water.

Almost twice as long as a great white shark, this ocean **monster** was one of the biggest animals to have ever swam in the water.

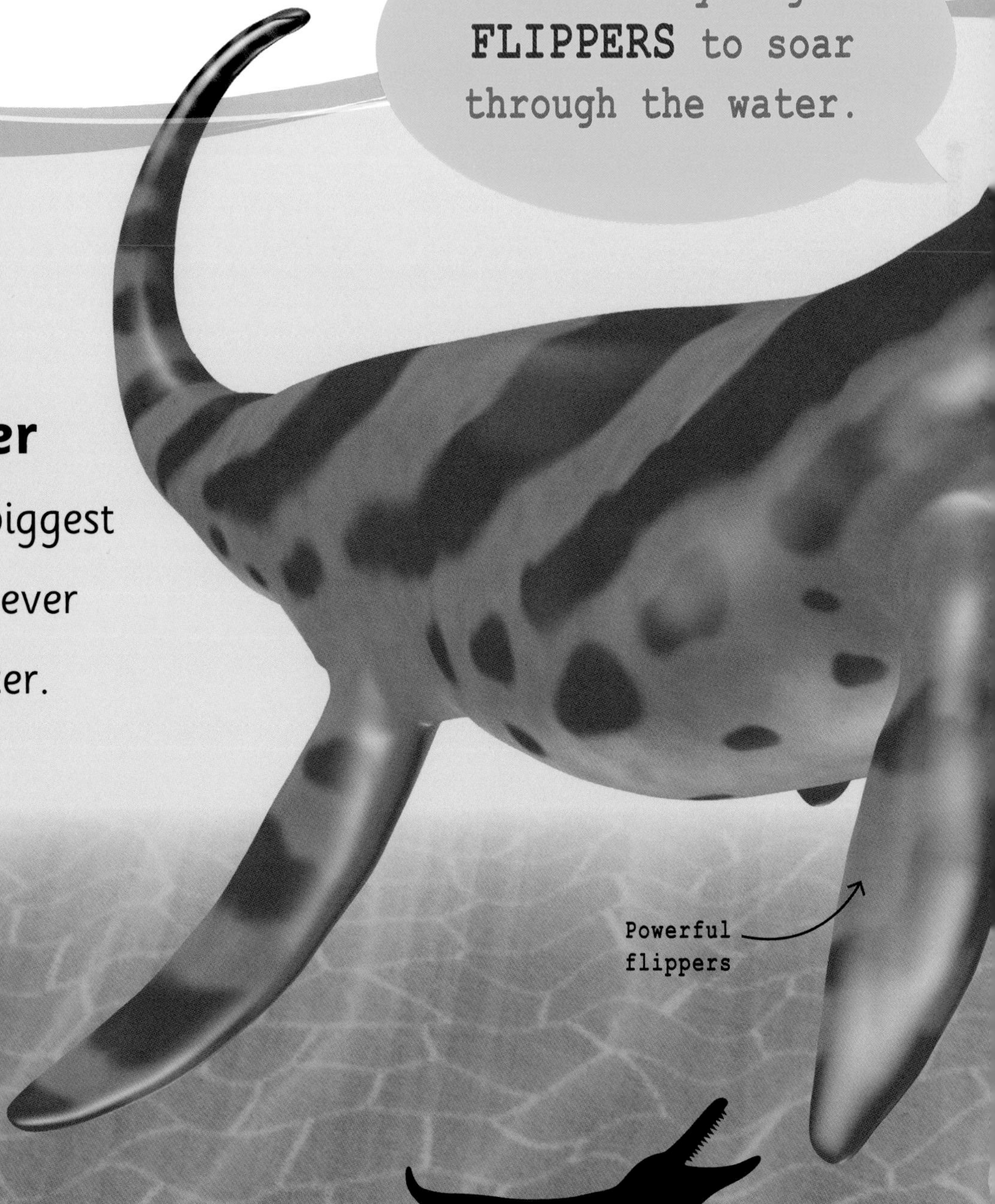

Powerful flippers

251 million years ago

200

145

66

Triassic

Jurassic

Cretaceous

Fact File

Speedy hunter

Kronosaurus' best weapon was its **huge snout**. Its head was 3m (10ft) long and its jaws opened wide, like a crocodile's. Its mouth was full of sharp, banana-sized teeth.

No fussy eater

Kronosaurus probably ate **whatever** it could catch. Kronosaurus fossils have been found that had other plesiosaurs and sea turtles in the stomach. It probably also ate fish and squid.

Size: 10m (33ft) long **Habitat:** Oceans **Diet:** Fish and marine reptiles

Clues from

the past

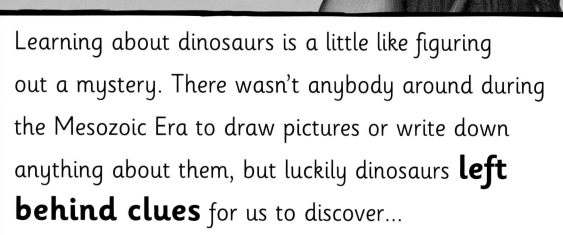

Learning about dinosaurs is a little like figuring out a mystery. There wasn't anybody around during the Mesozoic Era to draw pictures or write down anything about them, but luckily dinosaurs **left behind clues** for us to discover...

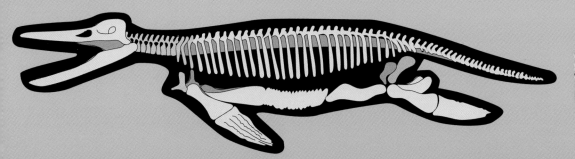

What are fossils?

How do we know anything about dinosaurs and other prehistoric creatures if they're no longer around? We study their **fossils**.

A fossil can show us if a dinosaur had wings or feathers.

What are they?

Fossils are the remains of living things that died a long time ago and have **been preserved** in rock, mud, sand, or gravel.

If a dinosaur had sharp teeth and long claws, it was probably (but not always) a meat-eater.

Teeth

The bigger its bones, the bigger the dinosaur was.

Bone

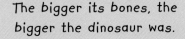

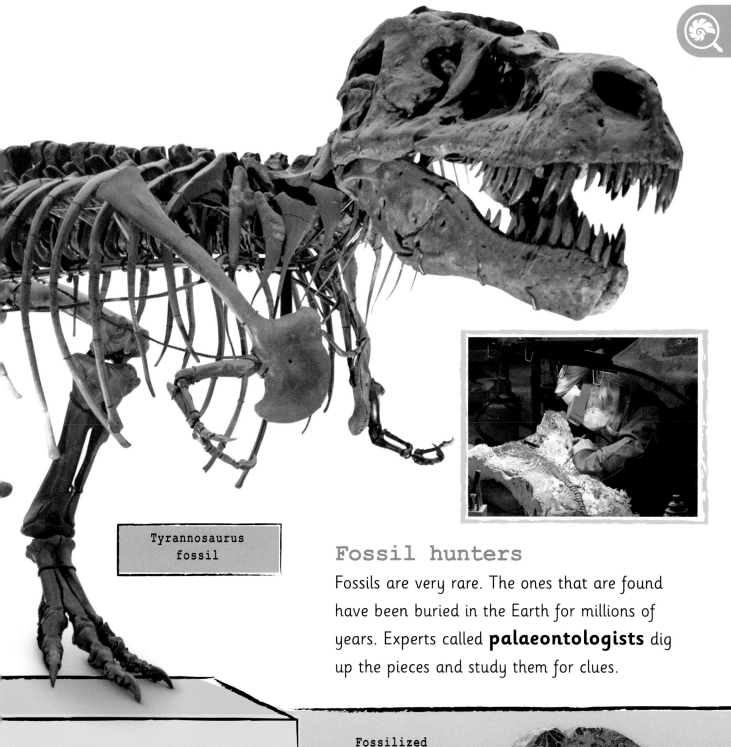

Tyrannosaurus
fossil

Fossil hunters

Fossils are very rare. The ones that are found have been buried in the Earth for millions of years. Experts called **palaeontologists** dig up the pieces and study them for clues.

Pieces of a puzzle

It's rare to find full dinosaur skeletons, so experts often look at bones from several skeletons to make a whole one. It's a bit like doing a very difficult **jigsaw** puzzle.

Fossilized
Ginkgo plant

Animals, plants, and other
living things can all
become fossils too.

Types of fossil

There are several types of fossil. Most form by **turning to stone**, but there are several other ways as well.

The process of something turning to stone is called PETRIFICATION.

Cast of a skull

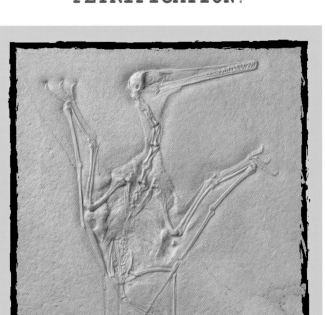

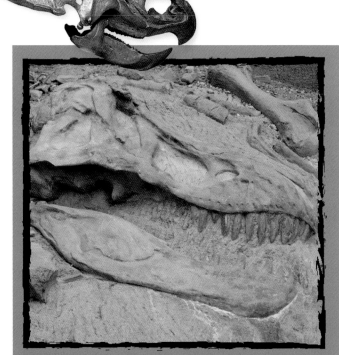

Sometimes the remains of an animal or plant will fade, but leave behind an impression **(mold)** of itself in rock or mud.

Casts start off like molds, but minerals in water fill them from the inside, eventually creating a copy of the original shape.

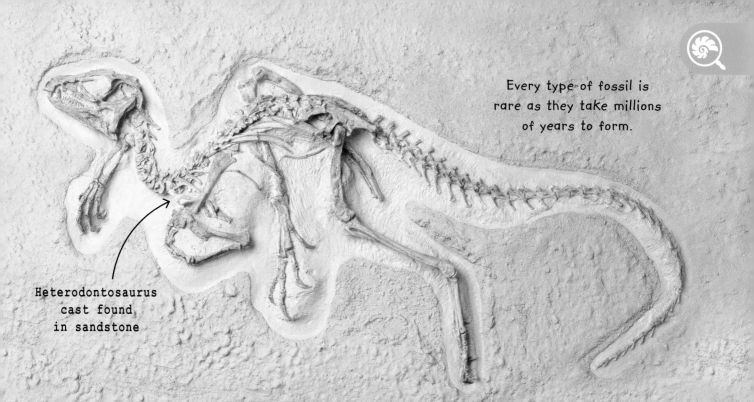

Heterodontosaurus cast found in sandstone

Insects or other small animals that became stuck in sticky tree sap have been found totally preserved in **amber** (fossilized tree resin).

Sometimes, all that remains of a creature is their footprints, teeth, or poo. These are known as **trace fossils**.

First fossil finds

For a very long time, people had **no idea** that dinosaurs ever existed. But that all changed when people began studying fossils.

A giant mystery

In 1677, when **Robert Plot** was working in a museum in Oxford, England, he came across a giant fossil bone. He compared the bone to different animals, but they were all the wrong shape and size.

This bone is huge! What kind of creature could it belong to?

Drawing of the mystery bone

Robert Plot thought the fossil was the bone of a giant person from ancient times, but we now know it was a leg bone of a Megalosaurus.

Megalosaurus

Interesting teeth

In 1820, **Gideon** and **Mary Ann Mantell** found tooth fossils that looked like an iguana's, but bigger. They named the newly discovered animal "Iguanodon".

Iguanodon teeth

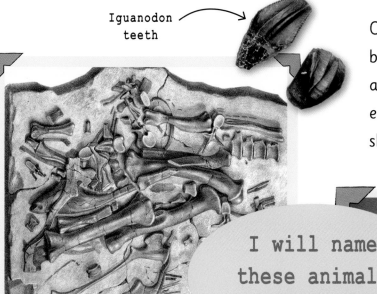

Other scientists thought the teeth belonged to a rhinoceros, but Mary and Gideon kept studying, and eventually found a full Iguanodon skeleton and one of a Hylaeosaurus.

> I will name these animals DINOSAURS!

The dinosaur family

In 1842, **Sir Richard Owen** realized the fossils of Megalosaurus, Iguanodon, and Hylaeosaurus belonged a **group** of extinct animals — dinosaurs!

It was Sir Richard Owen who came up with the name dinosaurs. It means "terrible lizards".

Mary Anning

Even though she was very poor and didn't go to school, Mary Anning made lots of discoveries and became one of the most famous and respected **fossil hunters** ever.

Her first discovery

In 1811, when Mary was just 12, she and her brother Joseph found a huge skull fossil. Mary found the rest of the skeleton and spent a year digging it out. It turned out to be the first **ichthyosaur** fossil ever discovered.

Ammonite

Plesiosaur fossil

Icthyosaur skull

PLESIOSAURUS MACROCEPHALUS.
LYME REGIS, DORSETSHIRE
LOWER LIAS

Lots of firsts

As well as the first ichthyosaur, Mary discovered the first **plesiosaur**, the first **pterosaur**, a giant prehistoric fish, ammonites, and other fossils that had never been seen before.

As a little girl, Mary sold shells to help her family.

LONDON

LYME REGIS

With the help of her dog, Mary looked for fossils on the beach at Lyme Regis in Dorset, England.

A hard life

Growing up in a poor family, Mary couldn't afford go to school. As an adult, she had to **sell** her fossils to earn money. She kept drawings and notes, but often didn't get credit for her discoveries.

Henry De la Beche painting

The artist Henry De la Beche painted the animals that Mary discovered to help raise money for her studies.

A world of fossils

Dinosaur fossils may be rare, but they have been found all over the world.

Struthiomimus found in Canada

Canada
A Struthiomimus skeleton found in Alberta, Canada in 1914 is one of the most complete dinosaur fossils ever discovered.

Diplodocus found in Wyoming, USA

USA
A near complete skeleton of a Diplodocus was discovered by railway workers in Wyoming in 1898.

USA
Thousands of dinosaur bones have been found at the "Dinosaur National Monument", near the border between Utah and Colorado.

Argentinosaurus found in Argentina

Argentina
The biggest dinosaur ever discovered, Argentinosaurus, was found in a desert near La Flecha, Argentina in 2014.

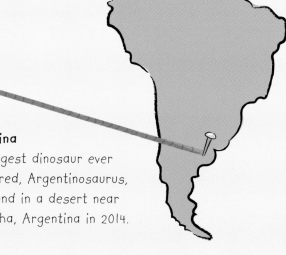

United Kingdom

Megalosaurus fossils found in Oxfordshire, England, inspired William Buckland to write the first scientific report about dinosaurs.

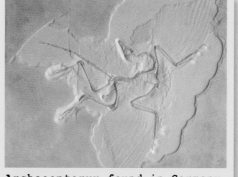

Archaeopteryx found in Germany

Germany

In 1860, Archaeopteryx was found in Germany. It turned out to be the missing link between dinosaurs and birds.

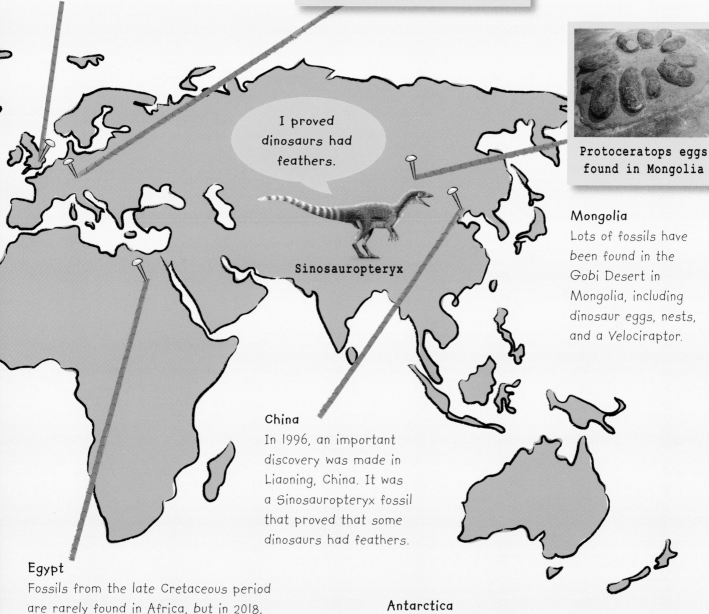

I proved dinosaurs had feathers.

Sinosauropteryx

Protoceratops eggs found in Mongolia

Mongolia

Lots of fossils have been found in the Gobi Desert in Mongolia, including dinosaur eggs, nests, and a Velociraptor.

China

In 1996, an important discovery was made in Liaoning, China. It was a Sinosauropteryx fossil that proved that some dinosaurs had feathers.

Egypt

Fossils from the late Cretaceous period are rarely found in Africa, but in 2018, a new sauropod was discovered in Egypt. It was named Mansourasaurus.

Antarctica

Dinosaur fossils have even been found on Antarctica. One example is Cryolophosaurus, a theropod from the early Jurassic.

How fossils are made

A lot of things need to happen for an animal or plant to become a fossil. It also takes **millions of years**. Here's what happens.

Some fossils have been found that are more than 3 billion years old!

Fossilization

Fossils only form if an animal or plant is **buried** quickly after death. Otherwise it will just decay and disappear. This is why fossils are so rare.

1 A dinosaur dies and becomes **buried** in thick mud.

2 Over a few years the flesh rots away, leaving just the **bones**.

3 Over millions of years, Earth's plates shift and a sea forms above the burial site. Slowly, the bones turn into **rock**.

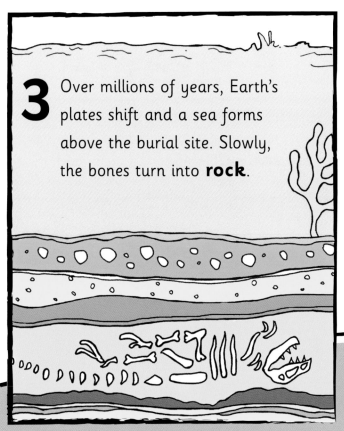

4 Earth's plates continue to shift until the sea eventually disappears, leaving the **ground** exposed.

Wind wears away the surface.

5 Years and years later, the layers of ground above the fossil wear away and the fossil is **discovered**. Finally, a team of skilled palaeontologists can begin to dig the fossil up!

From mud to museum

I was found buried in a swamp in China.

Museums are amazing places full of **fossils** and **prehistoric treasures**. But have you ever wondered how fossils get there?

1 A dinosaur dies and is buried in the Earth. Over millions of years, it becomes a fossil. Years later the fossil is discovered.

2 Palaeontologists start working to excavate (dig out) the fossil with tools. The fossil is coated in plaster to protect it.

3 The palaeontologists take detailed photographs and sketches of everything. These records will be very useful to study later on.

The excavation process can take several months.

4 The fossil is packed up and sent to a laboratory to be studied or to a museum for display.

Fossils are rare and delicate, so palaeontologists have to be **VERY CAREFUL** when they work with them.

Some bones are too heavy or delicate to display, so copies are sometimes made from fibreglass.

Palaeontologist tool kit

Digging up dinosaur bones is **slow, careful work**, so palaeontologists need a special selection of tools to help them uncover the delicate fossils.

Tools of the trade

Palaeontologists use very expensive high-tech equipment, but they also use **everyday objects** like cutlery and paintbrushes. Each dig needs special tools depending on the size of the site and how fragile the fossils are.

Camera

It's important to document the dig with lots of photos.

Walkie-talkies

These radios are perfect for keeping in touch with other members of the team.

Tooth picks

Wooden picks are great to clean out tiny, delicate things.

Selection of brushes

Brushes can gently sweep away dust. Toothbrushes are perfect for cleaning fossils.

Pen

Small bags

Small finds are placed in bags and labelled.

Tape measure

All the different parts of a fossil need to be measured.

A special solution helps stick pieces of a fossil together if they have cracked.

Solution

Notebook

Hand pick

Hand picks can remove rough stone from around a fossil.

Chisel and hammer

A chisel and hammer can chip away at hard mud.

Spoons

Spoons are used to carefully dig around fossils.

Trowel

Trowels make fast and precise digging tools.

Line pins

Pins and string are used to mark the dig area.

Ruler and pencil

Scientists need to take careful notes and make drawings of the site as they dig.

String

Feathered family

While dinosaurs as we think of them are gone, some feathered theropods survived the end of the Mesozoic Era and eventually became **birds**.

Early birds

Small theropods evolved over millions of years to become the birds we know today — making birds the only **surviving dinosaurs**!

Look at all my feathered friends.

161
million years ago

Anchiornis was feathered and would have been able to glide.

151
million years ago

Archaeopteryx was probably one of the first dinosaurs that could fly.

There are more than 10,000 different types of bird

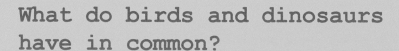

What do birds and dinosaurs have in common?

Scaly foot

Eggs

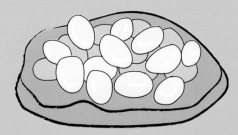

Dinosaurs and birds hatch from eggs. Lots of them made nests and some even sat on their eggs to keep them warm and safe.

Scales

If you look closely at a bird's legs, you will see that they are scaly, like reptiles' bodies.

Feathers

Lots of dinosaurs had feathers but not all could fly. Early feathers were soft, fluffy, and used to keep warm or to impress mates.

Feathered wing

125 million years ago

68 million years ago

56 million years ago

Iberomesornis was well built for flight. It had a little claw on its wings.

Vegavis was closely related to modern ducks and geese.

Gastornis was a giant, flightless bird with fluffy feathers like a kiwi bird's.

alive today. They are dinosaur's **LIVING RELATIVES.**

Where did the dinosaurs go?

Dinosaurs ruled the Earth for nearly 170 million years. Then, 66 million years ago a meteorite crashed into the planet, causing the dinosaurs to become **extinct** (die out).

Experts think the meteorite was about 10km (6 miles) wide!

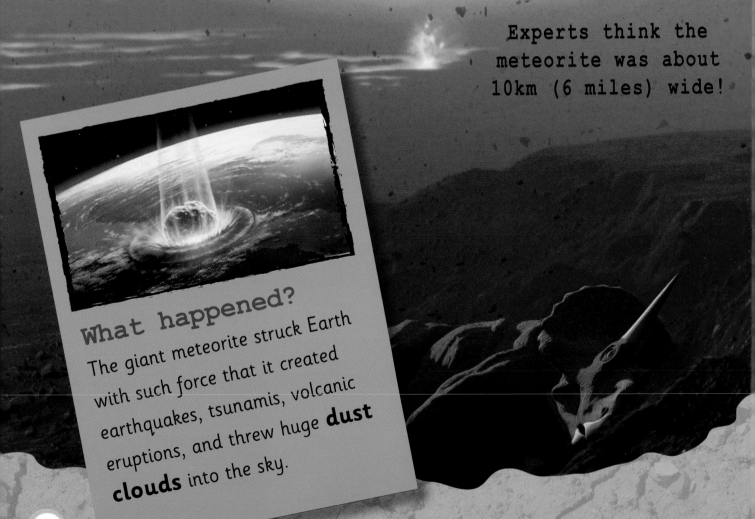

What happened?

The giant meteorite struck Earth with such force that it created earthquakes, tsunamis, volcanic eruptions, and threw huge **dust clouds** into the sky.

What happened next?

The dust clouds made it hard for animals to breathe and blocked out the Sun's light and heat. This caused the **Earth's climate to change**. As a result, there wasn't enough food to support life for most species.

What survived?

Only small land animals, as well as some **fish, lizards, and insects** were able to survive. Over millions of years, new animals came and went, until finally humans arrived on the scene.

New discoveries about the past

You might think that because dinosaurs lived so long ago, we already know everything about them. But experts are making **new discoveries** all the time.

Stuck in time

In 2016, a Chinese palaeontologist made a once in a lifetime discovery – a dinosaur tail trapped in **amber**. 100 million years earlier, a dinosaur became stuck in the resin and its tail bones, skin, blood, and feathers were totally preserved.

Tail feathers

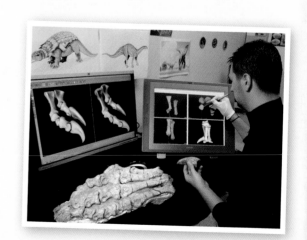

Sometimes, new technology such as scanners and lasers lets us look inside fossils and work out what colour the dinosaur would have been.

New (but old) animals

Scientists still discover **new dinosaurs**. The duck-like Halszkaraptor was discovered in 2017. It had curved claws, a swan-like neck, and limbs that may have let it swim. Mansourasaurus is a new type of sauropod found in Egypt in 2018.

Halszkaraptor

Halszkaraptor might have lived on land and in water like a duck.

Mansourasaurus

I had flat teeth and a long neck, so I probably ate like a giraffe.

What did dinosaurs do?

Fossils help us understand what dinosaurs **look like**, but they can only tell us a little about how dinosaurs behaved. For this, scientists **compare dinosaurs to modern animals** such as birds, giraffes, lizards, or crocodiles, and try to work things out.

Some discoveries teach us where we've been wrong. For years we had no idea dinosaurs could have feathers, but now we know they did.

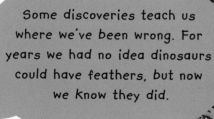

Dino file

Dinosaurs, pterosaurs, and plesiosaurs can have names that are hard to say. Check this guide if you need a little help.

● Triassic
○ Jurassic
● Cretaceous

Allosaurus
al-oh-SORE-uss

Amargasaurus
ah-MAR-gah-SORE-us

Anchiornis
AN-kye-OR-niss

Ankylosaurus
ANK-ill-oh-SORE-us

Apatosaurus
a-PAT-oh-SAW-russ

Archaeopteryx
ar-kee-OP-ter-ix

Argentinosaurus
ARE-jen-teen-oh-SORE-us

Brachiosaurus
brackee-oh-SORE-uss

Caudipteryx
kor-DIP-ter-iks

Chasmosaurus
KAZ-mo-SORE-us

Citipati
sih-tee-PA-tee

Coelophysis
SEE-lo-FYE-sis

Compsognathus
COMP-sog-NAITH-us

Corythosaurus
ko-RITH-oh-SORE-us

Cryolophosaurus
cry-o-LOAF-o-SORE-us

Deinonychus
dye-NON-ee-cuss

Dimorphodon
dye-MOR-foh-don

Diplodocus
dip-LOD-oh-kus

Edmontonia
ED-mon-TOE-nee-a

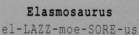

Edmontosaurus
ed-MONT-oh-SORE-us

Einiosaurus
ie-nee-oh-SAWR-uss

Elasmosaurus
el-LAZZ-moe-SORE-us

Eoraptor
EE-oh-RAP-tor

Eudimorphodon
YOU-die-MORE-fo-don

Euoplocephalus
YOU-owe-plo-SEFF-ah-luss

Gallimimus
GAL-ih-MIME-us

Giganotosaurus
gig-AN-oh-toe-SORE-rus

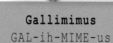

Guanlong
GWON-long

Halszkaraptor
hals-ka-RAP-tor

Hamipterus
ham-IP-ter-us

Herrerasaurus
heh-RARE-ra-SORE-uss

Heterodontosaurus
HET-er-oh-DON-toe-SORE-us

Huayangosaurus
hoy-YANG-oh-SORE-uss

Hypsilophodon
hip-sih-LOAF-oh-don

Ichthyosaurus
ICK-thee-oh-SORE-uss

Iguanodon
ig-WAH-no-don

Kentrosaurus
KEN-troh-SORE-uss

Kronosaurus
crow-no-SORE-us

Lambeosaurus
LAM-bee-oh-SORE-uss

Latenivenatrix
lah-ten-EYE-vuh-NAY-
tricks

Liopleurodon
LIE-oh-PLOOR-oh-don

Maiasaura
MY-ah-SORE-a

Mansourasaurus
man-SOO-rah-SORE-us

Megalosaurus
MEG-uh-lo-SORE-us

Microraptor
MY-crow-rap-tor

Mosasaurus
MOES-ah-saw-rus

Ornithocheirus
or-NITH-oh-KEE-rus

Ouranosaurus
oo-RAH-no-SORE-uss

Pachycephalosaurus
PACK-ee-sef-ah-low-SORE-us

Parasaurolophus
PA-ra-SORE-oh-LOAF-uss

Pentaceratops
PEN-ta-SERRA-tops

Plateosaurus
plate-ee-oh-SORE-us

Protoceratops
PRO-toe-SERRA-tops

Psittacosaurus
si-tak-ah-SORE-us

Pteranodon
teh-RAN-oh-don

Quetzalcoatlus
ket-zal-KWAT-luss

Rhamphorhynchus
ram-foe-RINK-us

Rhomaleosaurus
ROME-alley-oh-SORE-us

Sauropelta
SORE-oh-PELT-ah

Shunosaurus
SHOE-noe-SORE-us

Sinosauropteryx
SIGH-no-SORE-op-ter-ix

Spinosaurus
SPINE-oh-SORE-us

Stegosaurus
STEG-oh-SORE-uss

Struthiomimus
STROO-thee-oh-
MIME-us

Styracosaurus
sty-RACK-oh-SORE-us

Suchomimus
SOO-ko-MIME-us

Thecodontosaurus
THEE-co-DON-toe-SORE-us

Therizinosaurus
THERRY-zin-oh-SORE-us

Torosaurus
TOH-row-SAW-russ

Triceratops
try-SERRA-tops

Tsintaosaurus
SIN-tow-SORE-uss

Tupandactylus
too-pan-DAK-til-us

Tyrannosaurus
tie-RAN-oh-SORE-us

Utahraptor
YOU-tah-RAP-tor

Velociraptor
vel-OSS-ee-rap-tor

Dinosaur words

This book is filled with big dinosaur words. Some can be a bit tricky, so if you ever get stuck, look here.

Amphibian A group of animals that can live on both land and in water.

Armour Features that protect animals from harm, such as spikes and plates.

Carnivore An animal that eats other animals.

Ceratopsian A subgroup of thyreophorans.

Climate The weather in a place or during a period of time.

Cretaceous The third period of the Mesozoic Era.

Dromaeosaur A subgroup of theropods.

Environment The physical surroundings of a living thing.

Evolve The way living things change over time to help them survive.

Extinct When a group of animals or plants completely die out.

Fossil The remains of living things that died a long time ago and have been preserved in the Earth.

Habitat The natural environment of an animal.

Hadrosaur A subgroup of ornithopods.

Herbivore An animal that only eats plants.

Herds A pack of animals that live or travel together.

Jurassic The second period of the Mesozoic Era.

Mammal Warm-blooded animals that drink their mother's milk as babies.

Marginocephalian A group of dinosaurs that had bony frills on their heads.

Mesozoic Era The time dinosaurs existed. Made up of three periods called the Triassic, Jurassic, and Cretaceous.

Museum A place that displays historical artifacts.

Omnivore Animals that eat plants as well as meat.

Ornithopod A group of dinosaurs that grazed on plants and sometimes travelled in herds.

Palaeontologist A scientist who studies fossils and prehistoric life.

Pangaea The C-shaped supercontinent that made up Earth's land at the beginning of the Mesozoic.

Plesiosaur Prehistoric sea-dwelling reptiles.

Predator An animal that hunts and kills other animals for food.

Prey Animals that are eaten by predators.

Pterosaurs Prehistoric flying reptiles.

Reptile A group of cold-blooded, scaly animals.

Sauropodomorph A group of mostly very tall, plant-eating dinosaurs.

Sauropods Huge plant-eating dinosaurs with long necks and tails.

Skeleton A frame of bones that supports an animal's body.

Species A group of similar animals that share the same features and can have babies together.

Spine An animal's backbone.

Tectonic plates Large sections of rock beneath the Earth's surface, that move very slowly.

Theropod A group of meat-eating dinosaurs that walked on two legs.

Thyreophoran A group of plant-eating dinosaurs that were heavily protected with armour.

Triassic The first period of the Mesozoic Era.

Index

Acknowledgements

The publisher would like to thank the following for their kind permission to reproduce their photographs:

Key: a= above; b=below/bottom; c=centre; f=far; l=left, r=right, t=top.

1 123RF.com: Linda Bucklin (bl, br); Corey A Ford (tr, tc); Elena Duvernay (crb). **2 123RF.com:** Linda Bucklin (cra); Corey A Ford (tr). **Dorling Kindersley:** Tim Ridley / Robert L. Braun (bc). **Getty Images:** Nobumichi Tamura / Stocktrek Images (bl). **3 123RF.com:** chastity (bl). **4 123RF.com:** Corey A Ford (bc, tc, tr). **5 123RF.com:** Linda Bucklin (br); Michael Rosskothen (bc). **iStockphoto.com:** Corey Ford (cr). **6 123RF. com:** Corey A Ford (bl). **Dorling Kindersley:** Peter Minister (br). **7 123RF.com:** Corey A Ford (br). **8 123RF.com:** Linda Bucklin (tl, cr); Valentyna Chukhlyebova (tr). **Dorling Kindersley:** Tim Ridley / Robert L. Braun (bl). **9 123RF.com:** Corey A Ford (br). **Alamy Stock Photo:** Science Photo Library (t). **10-11 123RF.com:** Corey A Ford (t). **11 123RF.com:** Simone Gatterwe (bc); Michael Rosskothen (br). **12 123RF.com:** Michael Rosskothen (bl). **James Kuether:** (c, cr). **13 123RF.com:** leonello calvetti (c); Michael Rosskothen (cb). **Dorling Kindersley:** Andy Crawford / Royal Tyrrell Museum of Palaeontology, Alberta, Canada (br). **14-15 123RF.com:** Nataliya Hora (background); pakhnyushchyy (t/background). **14 123RF.com:** Michael Rosskothen (br). **Dorling Kindersley:** Jon Hughes (cr). **15 123RF.com:** Michael Rosskothen (c). **Dorling Kindersley:** Colin Keates / Natural History Museum, London (br). **16 123RF.com:** Linda Bucklin (bl). **Alamy Stock Photo:** leonello calvetti (c); Science Photo Library (cr). **16-17 123RF.com:** Ievgenii Biletskyi (c/background); pakhnyushchyy (t/background); daveallenphoto (cb/background). **17 123RF.com:** Corey A Ford (tc); Michael Rosskothen (tl). **Dreamstime.com:** Vaeenma (br). **James Kuether:** (c). **18 123RF.com:** leonello calvetti (cb); Corey A Ford (cr, ca). **18-19 123RF.com:** Alberto Loyo (c/background). **19 123RF.com:** leonello calvetti (bl, t). **123RF.com:** Tim Hester / timhester (br). **20-21 123RF.com:** marina gallud (b/background). **21 Dorling Kindersley:** Frank Greenaway / Natural History Museum (c). **22 123RF.com:** Mariusz Jurgielewicz (bl); yobro10 (cr) **Dreamstime.com:** Pablo Hidalgo / Pxhidalgo (cl). **22-23 Dreamstime.com:** Yellowdesignstudio (masking tape). **23 123RF.com:** leonello calvetti (b); sborisov (t). **24 123RF.com:** leonello calvetti (cr); Mark Turner (cra); Valentyna Chukhlyebova (cl). **25 123RF.com:** leonello calvetti (cb); Mark Turner (cla); Corey A Ford (bc, c, clb). **26 123RF.com:** Linda Bucklin (tr). **James Kuether:** (tc). **27 James Kuether:** (bl, br). **28 Alamy Stock Photo:** Stocktrek Images, Inc. (cr). **123RF.com:** Teresa Gueck / teekaygee (cl). **29 123RF.com:** Suman Bhaumik (tr). **30 123RF.com:** leonello calvetti (r). **31 123RF.com:** Elena Duvernay (l); Mark Turner (c). **32 123RF.com:** Aliaksei Hintau / viselchak (cra); Dreamstime.com: Vladimirdavydov (br); Dreamstime.com: Vivid Pixels / Vividpixels (tr); Dreamstime.com: Alle (cb, crb). **33 123RF.com:** rodho (t). **123RF.com:** Alexandr Pakhnyushchyy / alekss (clb); Dreamstime.com: Sutisa Kangvansap / Mathisa (br); 123RF.com: Brandon Alms / macropixel (bl); Frank Greenaway / Natural History Museum, London (cra); Forrest L. Mitchell / James Laswel (cl). **34 123RF.com:** Corey A Ford (cr). **Alamy Stock Photo:** Science Photo Library (ca). **35 123RF.com:** Corey A Ford (c). **Dreamstime.com:** Mary (cr); Gary Ombler / Senckenberg Gesellschaft Fuer Naturforschugn Museum (b). **36 123RF.com:** Mark Turner (cl). **37 Dorling Kindersley:** Jon Hughes (cr). **38 123RF.com:** citadelle (c). **Dorling Kindersley:** Peter Minister and Andrew Kerr / Dreamstime.com: (cb); Harry Taylor / Natural History Museum, London (b). **38-39 Dreamstime.com:** Kelvintt (t/background). **39 123RF.com:** citadelle (cb). **Alamy Stock Photo:** Stocktrek Images, Inc. (br). **40 123RF.com:** Michael Rosskothen (br). **41 123RF.com:** Corey A Ford (tr); Michael Rosskothen (cl). **42 123RF.com:** dirkr (cl); Dreamstime.com: Nataliya Hora / Zhu_zhu (b/background); 123RF.com: Vitalii Gulay / vitalisg (br). **43 123RF.com:** Corey A Ford (bl); Tamara Kulikova (bc). **James Kuether** (br). **44 123RF.com:** Vasyl Hedzun (cr); suwat wongkham (tr). **Dreamstime.com:** Glenda Powers / Mcininch (bl). **James Kuether** (cl, br). **45 123RF.com:** Elena Duvernay (t). **James Kuether** (c, b). **46 Dreamstime.com:** Aleksandr Frolov / Afhunta (clb); Colin Keates / Natural History Museum (tl, bl, br, cr). **47 James Kuether** (b). **48 123RF.com:** Mariusz Blach (cr); gkuna (br); Corey A Ford (c). **49 123RF.com:** leonello calvetti (c); Nico Smit (tr). **Alamy Stock Photo:** Science Photo Library (tl). **50 123RF. com:** Camilo Maranchón garcía (br). **50-51 123RF.com:** Corey A Ford (t). **51 123RF. com:** Michael Rosskothen (bl). **52 123RF.com:** Valentyna Chukhlyebova (br). **Dreamstime.com:** Corey A. Ford / Coreyford (bl). **53 123RF.com:** Linda Bucklin (br). **Nobumichi Tamura:** (bl). **54 123RF.com:** Simone Gatterwe (cl); Michael Rosskothen (c). **Dreamstime.com:** Witold Krasowski / Witoldkr1 (br); Dorling Kindersley / Peter Minister (cr). **55 123RF.com:** Corey A Ford (br). **Dreamstime:** Boborsillo (tr). **56 123RF. com:** Michael Rosskothen (bc). **56-57 James Kuether** (t). **58 Alamy Stock Photo:** Xinhua (c). **Dorling Kindersley:** John Downes / Natural History Museum, London (crb). **59 123RF.com:** Corey A Ford (br). **60 123RF.com:** leonello calvetti (bl); Corey A Ford (t). **61 123RF.com:** Mark Turner (t). **Alamy Stock Photo:** leonello calvetti (bl); MasPix (br). **62-63 123RF.com:** Lorelyn Medina (b). **64-65 123RF.com:** Tommaso Lizzul (background). **65 Dorling Kindersley:** Andy Crawford / Royal Tyrrell Museum of Palaeontology, Alberta, Canada (b). **66 Depositphotos Inc:** Aliencat (bl). **Dreamstime.com:** Prapass Wannapinij / Prapass (cr). **66-67 123RF.com:** Elena Duvernay (background). **67 123RF.com:** softlight69 (t/background). **Depositphotos Inc:** Aliencat (c). **Dorling Kindersley:** Gary Ombler / Senckenberg Gesellshaft Fuer Naturforschugn Museum (b). **68-69 Dreamstime.com:** Inga Nielsen / Ingan (background). **iStockphoto.com:** Elenarts (c). **68 Dorling Kindersley:** Colin Keates / Natural History Museum, London (bl). **69 Dorling Kindersley:** Corey A Ford (b). **70-71 123RF. com:** bazru (background). **Dreamstime.com:** Valentin Armianu / Asterixvs (t/background). **72 123RF.com:** Corey A Ford. **72-73 123RF.com:** daveallenphoto (background). **73 123RF.com:** Corey A Ford (b). **Science Photo Library:** Natural History Museum (tr). **74 Alamy Stock Photo:** Corey Ford (cb). **75 James Kuether. 76 Dorling Kindersley:** Andy Crawford / Royal Tyrrell Museum of Palaeontology, Alberta, Canada (br). **77 123RF.com:** Corey A Ford (c). **78-79 123RF.com:** Lorelyn Medina (b). **Alamy Stock Photo:** leonello calvetti (c). **Dreamstime.com:** Inga Nielsen / Ingan (t/background). **82-83 Getty Images:** Yuriy Priymak / Stocktrek Images. **83 123RF.com:** W.Scott McGill (tr). **Wikimedia:** Yuya Tamai / CC BY 2.0 (br). **84-85 123RF.com:** Michael Rosskothen. **84 Dorling Kindersley:** Andy Crawford / State Museum of Nature, Stuttgart (br). **86 123RF.com:** Corey A Ford (c); Michael Rosskothen. **Wikimedia:** Nkansahrexford / CC BY 3.0 (b). **87 123RF.com:** softlight69 (background). **88-89 123RF.com:** Mr.Smith Chetanachan (b/background); Nataliia Kravchuk (t/background). **88 123RF.com:** chastity (cl). **89 123RF.com:** Mark Turner. **90 123RF. com:** Simone Gatterwe (cl); Michael Rosskothen (cr). **91 123RF.com:** Michael Rosskothen; pongbun sangkaew (t/background). **92 123RF.com:** softlight69 (t/background). **93 123RF.com:** Shlomo Polonsky (br). **94-95 123RF.com:** Keith Halterman (b/background). **94 James Kuether. 95 Dorling Kindersley:** Miguel Periera / Museo Arentino De Cirendas (br). **James Kuether** (tl). **96-97 123RF.com:** sebastien decoret (background); sborisov (b/background). **96 123RF.com:** leonello calvetti; netsuthep summat (bl). **97 123RF.com:** leonello calvetti (c). **98-99 123RF.com:** Lorelyn Medina (b). **100-101 123RF.com:** Michael Rosskothen (b). **123RF.com:** szefei (background). **100 123RF.com:** Lefteris Papaulakis (l). **102 Dorling Kindersley:** Colin Keates / Natural History Museum, London (bl). **103 Dreamstime.com:** Leonello Calvetti / Leocalvett (br). **104-105 123RF.com:** Nataliia Kravchuk (t); Shlomo Polonsky (b/background). **James Kuether** (c). **105 James Kuether** (tr, cb). **108-109 123RF.com:** Shlomo Polonsky (b/background); Nico Smit (background). **109 123RF.com:** memoangeles (br). **Alamy Stock Photo:** MasPix (cb). **110 Dorling Kindersley:** Peter Minister (bl). **112-113 123RF.com:** Michael Rosskothen (b). **112 123RF.com:** Michael Rosskothen (br). **114 123RF.com:** Andreas Meyer (cl). **114-115 123RF.com:** Olga Khoroshunova (b/background); Andreas Meyer (c); Laurin Rinder (t/background). **115 123RF.com:** Corey A Ford (b); Michael Rosskothen (tr). **116-117 123RF.com:** Lorelyn Medina (b). **118-119 123RF.com:** Elena Duvernay. **120 James Kuether** (l). **121 123RF.com:** Linda Bucklin (cra); Elena Duvernay (cla). **James Kuether** (cl). **122-123 123RF.com:** Frederik

Johannes Thirion (background). **124-125 123RF.com:** Tim Hester / timhester (t/ background). **125 Getty Images:** Bernard Weil / Toronto Star (r, cra). **126 Science Photo Library:** Martin Shields (l). **127 Science Photo Library:** Julius T Csotonyi. **128 123RF.com:** Michael Rosskothen (cr). **Dreamstime.com:** Jinfeng Zhang (bl). **128-129 123RF.com:** szefei (background). **129 123RF.com:** Corey A Ford (c). **130-131 123RF.com:** Lorelyn Medina (b). **132 123RF.com:** iimages (bl). **Dorling Kindersley:** Gary Ombler / Senckenberg Gesellschaft Fuer Naturforschugn Museum (br). **132-133 Alamy Stock Photo:** Mohamad Haghani. **134 123RF.com:** leonello calvetti (bl, b). **Dorling Kindersley:** Colin Keates / Natural History Museum, London (br). **134-135 123RF.com:** Mariusz Blach (background). **136 123RF.com:** leonello calvetti (b). **136-137 123RF.com:** Corey A Ford; Nataliya Hora (background). **139 123RF.com:** Elena Duvernay (cb). **140-141 123RF.com:** Michael Rosskothen. **140 Dorling Kindersley:** Lynton Gardiner / American Museum of Natural History (cl). **142 Courtesy Vladislav Konstantinov.** **143 123RF.com:** Elena Duvernay (br). **144-145 123RF.com:** Ajay Bhaskar (b/background). **146 123RF.com:** Michael Rosskothen. **147 123RF.com:** Michael Rosskothen (c). **148-149 123RF.com:** Lorelyn Medina (b). **150 Dorling Kindersley:** Andy Crawford / Royal Tyrrell Museum of Palaeontology, Alberta, Canada (cl). **150-151 123RF.com:** Linda Bucklin (c); Vassiliy Prikhodko (t). **151 123RF.com:** Linda Bucklin (br). **152-153 Alamy Stock Photo:** Stocktrek Images, Inc. **153 Alamy Stock Photo:** Riccardo Bianchini (tr). **154-155 123RF.com:** softlight69 (background). **Getty Images:** Nobumichi Tamura / Stocktrek Images. **156-157 123RF.com:** softlight69 (background). **158 123RF.com:** Elena Duvernay (cl). **Dorling Kindersley:** Lynton Gardiner / American Museum of Natural History (b). **161 123RF.com:** Shlomo Polonsky (r). **162-163 123RF.com:** Nico Smit (background). **James Kuether** (c). **163 123RF.com:** Alexey Sholom (br). **164 Getty Images:** Sergey Krasovskiy (cr). **164-165 Dreamstime.com:** Imagin.gr Photography (b/background). **165 Getty Images:** Sergey Krasovskiy (cra, b). **166-167 123RF.com:** Lorelyn Medina (b). **168 123RF.com:** Mark Turner (c). **169 123RF.com:** Corey A Ford (t). **170 123RF.com:** Corey A Ford (t). **170-171 123RF.com:** Eero Oskari Porkka (background); Mark Turner (c). **171 123RF.com:** Linda Bucklin (tr, cl); Andreas Meyer (c). **172 Alamy Stock Photo:** Xinhua (r, l). **173 Alamy Stock Photo:** Xinhua. **174 123RF.com:** Corey A Ford. **175 Dorling Kindersley:** Andy Crawford / Senckenberg Gesellschaft Fuer Naturforschugn Museum, (r); Jon Hughes (c). **176-177 123RF.com:** Michael Rosskothen (c). **Dreamstime.com:** Ulkass (t/background). **176 123RF.com:** Michael Rosskothen (tr, cra, cr). **177 123RF.com:** Corey A Ford (bl); Michael Rosskothen (tl). **Wikimedia:** Frank Kovalchek / CC BY 2.0 (r). **180-181 123RF.com:** Olga Khoroshunova (background); Shlomo Polonsky (b). **Alamy Stock Photo:** Science Photo Library (tr). **181 123RF.com:** Michael Rosskothen (tr). **182-183 123RF.com:** Michael Rosskothen (c). **182 Dorling Kindersley:** Jon Hughes (cr). **183 Dreamstime.com:** Jaroslav Moravcik (b). **184-185 123RF.com:** Corey A Ford. **184 Dorling Kindersley:** Colin Keates / Natural History Museum, London (b). **186-187 James Kuether.** **187 James Kuether.** **188-189 123RF.com:** Corey A Ford (c); Vassiliy Prikhodko (t/background). **189 123RF.com:** Corey A Ford (cr). **190 123RF.com:** Ian Dikhtiar (bl). **Dr Lida XING:** (tl). **191 Alamy Stock Photo:** Natural History Museum, London (cr). **Dorling Kindersley:** Colin Keates / Natural History Museum, London (tr). **Getty Images:** Jose Angel Astor (bc). **Dorling Kindersley:** Andy Crawford / State Museum of Nature, Stuttgart (clb); Lynton Gardiner / American Museum of Natural History (br); Colin Keates / Natural History Museum, London (bl). **Science Photo Library:** Millard H. Sharp (cl). **192-193 123RF.com:** Ian Dikhtiar (c). **193 Dorling Kindersley:** Gary Ombler / Swedish Museum of Natural History (br). **Getty Images:** benedek (cr). **194 123RF.com:** albertus engbers (cr). **Dorling Kindersley:** Colin Keates / Natural History Museum, London (cl, t). **195 123RF.com:** Sayompu Chamnankit (cr). **Dorling Kindersley:** Andy Crawford / Royal Tyrrell Museum of Palaeontology, Alberta, Canada (t); Tim Parmenter / Natural History Museum (cl). **196 Alamy Stock Photo:** Science History Images (bl); Science Photo Library (br). **Science Photo Library:** Paul D Stewart (tr). **Wellcome Images http://creativecommons.org/licenses/by/4.0/:** (tl). **197**

Alamy Stock Photo: Science History Images (tl, tr). **Dorling Kindersley:** Colin Keates / Natural History Museum, London (ca). **Science Photo Library:** Royal Institution of Great Britain (br). **Wikimedia:** (l). **198 Alamy Stock Photo:** Natural History Museum, London (tl; Natural History Museum, London (bl). **199 Alamy Stock Photo:** Natural History Museum, London (bl). **Dorling Kindersley:** Colin Keates / Natural (br). **200 123RF.com:** nattawat khodkaeo (bl); victor10947 (cl). **Dorling Kindersley:** Andy Crawford / Royal Tyrrell Museum of Palaeontology, Alberta, Canada (t). **201 123RF.com:** Narongrit Dantragoon (cr). **204 Getty Images:** Marius Hepp / EyeEm (l). **Science Photo Library:** Marco Ansaloni (c); Philippe Psaila (cr). **205 Dorling Kindersley:** Lynton Gardiner / American Museum of Natural History. **206 123RF.com:** grafner (bl); Vladimir Jotov (br); Anton Samsonov (tr). **206-207 123RF.com:** marina gallud (background). **207 123RF.com:** koosen (tl). **Dreamstime.com:** Ilya Genkin / Igenkin (ca). **208 123RF.com:** Corey A Ford (r). **209 123RF.com:** anat chantrakool (tr); Arunsri Futemwong (tc). **210 Science Photo Library:** David A Hardy (bl). **210-211 123RF.com:** Aleksandr Frolov (b/background). **Alamy Stock Photo:** Science Photo Library (t). **211 123RF.com:** Corey A Ford (clb, bl). **212 Science Photo Library:** Pascal Goetgheluck (br). **Dr Lida XING:** (bl). **213 123RF.com:** Corey A Ford (br); Subin pumsom (bc). **Carnegie Museum of Natural History:** Andrew McAfee (ca). **Malvit:** (tl). **214 123RF.com:** Elena Duvernay (cb/Caudipteryx); Corey A Ford (c/ Archaeopteryx, tc, cr/Chasmosaurus); Michael Rosskothen (bl, cr/Argentinosaurus). **Alamy Stock Photo:** leonello calvetti (tl; Science Photo Library (cl/Apatosaurus). **James Kuether** (br/Cryolophosaurus; bc/Compsognathus). **215 123RF.com:** Corey A Ford (bc/Hypsilophodon); Michael Rosskothen (tc/Dimorphodon, tl, bc/Huayangosaurus, cr/Giganotosaurus); Andreas Meyer (cr/Elasmosaurus). **Alamy Stock Photo:** Mohamad Haghani (cl/Euoplocephalus); Xinhua (cr/Hamipterus); MasPix (br). **James Kuether** (cla/ Edmontosaurus, cr/Gallimimus). **Malvit:** (cl/Halszkaraptor). **216 123RF.com:** Linda Bucklin (tr); Corey A Ford (cl/Ouranosaurus, bl, tl, cr/Maiasaura , tc/Kronosaurus, cb/ Microraptor); Mark Turner (cb/Parasaurolophus, br); Valentyna Chukhlyebova (tc/ Kentrosaurus); Michael Rosskothen (crb/Pentaceratops, cl/Liopleurodon). **Alamy Stock Photo:** Science Photo Library (cr/Mosasaurus). **Carnegie Museum of Natural History:** Andrew McAfee (cr/Mansourasaurus). **Depositphotos Inc:** Aliencat (bc/ Psittacosaurus). **Dorling Kindersley. 217 123RF.com:** Linda Bucklin (br); leonello calvetti (bc/Tyrannosaurus, cb/Triceratops); Mark Turner (bc/Utahraptor); Michael Rosskothen (cr/Suchomimus, cl/Styracosaurus). **Dorling Kindersley:** Jon Hughes (cr/ Thecodontosaurus). **Getty Images:** Sergey Krasovskiy (bl); Nobumichi Tamura / Stocktrek Images (crb). **James Kuether** (tc/Rhomaleosaurus, Shunosaurus). **218 123RF.com:** Mark Turner (bl). **Dorling Kindersley:** Colin Keates / Natural History Museum (t). **219 123RF.com:** Linda Bucklin (crb); Elena Duvernay (bl); Mark Turner (t); Corey A Ford (bc); Michael Rosskothen (br). **220 123RF.com:** Michael Rosskothen (crb). **221 123RF.com:** leonello calvetti (bl). **222 Getty Images:** Sergey Krasovskiy (cr). **224 123RF.com:** Michael Rosskothen (bc).

Cover images: *Front:* **123RF.com:** Linda Bucklin tr; *Back:* **123RF.com:** Corey A Ford cla; **Dorling Kindersley:** Tim Ridley / Robert L. Braun tc; **Getty Images:** Sergey Krasovskiy (cr).

All other images © Dorling Kindersley
For further information see: www.dkimages.com

DK would like to thank:
Satu Fox for proofreading and editorial assistance, Eleanor Bates and Kitty Glavin for design assistance, Marie Lorimer for indexing, Claire Cordier and Romaine Werblow for picture library assistance, Jo Walton for picture research, James Kuether for use of his artworks, and Vijay Kandwal, Pankaj Sharma, Sachin Singh, and Rajesh Singh Adhikari for digital editing.